匠艺整洁之道

程序员的职业修养

（英文版）

[美] 罗伯特 · C. 马丁 著

電子工業出版社
Publishing House of Electronics Industry
北京•BEIJING

内 容 简 介

鲍勃大叔因在技术人群中声名远播的 *Clean* 系列在全球圈粉无数。本书为其烫手新作，旨在为广大工程师指明一条通向匠师之路，包括饱经洗砺的敏捷技术实践，如何通过持续的努力提高专业素养，软件可用之上的目标与技能，以及如何激发团队最大潜能，等等。本书融会了几本经典著作的精髓，将“整洁”方法论推向至高境界——软件开发者有责任维护世界正常运行，而“人”才是“技术”的决定者。

本书共分三部分，前两部分用实例阐释 TDD 在敏捷软件中的运用，以及验收测试、协同编程等常被忽视的敏捷侧面与具体策略，还探讨了颇有价值的软件设计方案相关话题；第III部分拔地而起，直接提出十条堪称金玉良言的“规劝”，以帮助程序员成为团队基石。

本书适合所有软件开发者、测试工程师及工程类院校师生阅读，对技术团队负责人及架构师同样大有助益。

版权贸易合同登记号 图字：01-2022-0926

图书在版编目（CIP）数据

匠艺整洁之道：程序员的职业修养 = Clean Craftsmanship: Disciplines, Standards, and Ethics：英文 /（美）罗伯特 •C. 马丁著. —北京：电子工业出版社，2023.5

ISBN 978-7-121-44582-8

Ⅰ. ①匠… Ⅱ. ①罗… Ⅲ. ①软件开发—英文②程序设计—工程技术人员—职业道德—英文 Ⅳ.①TP311.52②G316

中国版本图书馆 CIP 数据核字（2022）第 222029 号

责任编辑：符隆美

印　　刷：三河市君旺印务有限公司

装　　订：三河市君旺印务有限公司

出版发行：电子工业出版社

　　　　　北京市海淀区万寿路 173 信箱　邮编：100036

开　　本：787×980　1/16　印张：25　　字数：508.4 千字

版　　次：2023 年 5 月第 1 版

印　　次：2023 年 5 月第 1 次印刷

定　　价：148.00 元

凡所购买电子工业出版社图书有缺损问题，请向购买书店调换。若书店售缺，请与本社发行部联系，联系及邮购电话：（010）88254888，88258888。

质量投诉请发邮件至 zlts@phei.com.cn，盗版侵权举报请发邮件至 dbqq@phei.com.cn。

本书咨询联系方式：（010）51260888-819，faq@phei.com.cn。

纪念迈克·比多（Mike Beedle）

专家推荐

鲍勃的《匠艺整洁之道》阐释了敏捷技术实践的目的，深入探讨了敏捷技术实践出现的历史因素，指出了敏捷技术实践为何总是那么重要。作者曾亲历敏捷技术的发展和成型过程，全面了解其实践目标和手段，这在本书中体现得淋漓尽致。

——蒂姆·奥廷格（Tim Ottinger）
知名敏捷教练，图书作者

鲍勃文风上佳。书稿易于阅读，概念解释得非常详尽，即便是新入行的程序员也能读懂。鲍勃也会时不时幽上一默，让你稍做放松。本书的真正价值在于呼唤变革，呼唤更好的东西……呼唤程序员的专业素养……以及对软件无处不在的认识。此外，我相信，鲍勃写到的历史还有很多价值。我很高兴地看到，他没有浪费时间指责我们如何走到今天。鲍勃呼吁大家行动起来，要求大家提高标准和专业素养，从而承担责任，即便有时这意味着某种退步。

——海瑟·坎瑟（Heather Kanser）

作为软件开发者，我们必须不断为雇主、客户、同事和未来解决重要问题。让软件可用尽管困难，但远未足够，并不能令你成为成功匠人。软件能运行，只代表你通过了能力测试。你也许具备成为匠人的能力，但还要掌握更多东西。在本书中，鲍勃阐明了能力测试之外的技能和责任，展示了严肃软件匠人该有的样子。

——詹姆斯·葛莱宁（James Grenning）

《测试驱动的嵌入式 C 语言开发》（*Test-Driven Development for Embedded C*）作者，

《敏捷宣言》（*Agile Manifesto*）作者之一

鲍勃是少数我愿意与之合作技术项目的知名开发者之一。并不只因为他技能出众、名声在外、善于沟通，更在于他曾帮助我成为更好的开发者和团队成员。他往往早于其他人好几年发现软件开发领域的重要变化趋势，且能解释其重要性，鼓舞我学习新技能。回顾我入行之时，匠艺和职业操守的概念还没在软件领域出现，大家只是告诉你要做个有诚信的好人。如今，这些概念已然成为专业开发者能习得的最重要能力，甚至比编码本身更为重要。我很高兴地看到鲍勃再领风气之先，迫不及待想听他阐述观点，并将他的观点应用于实践。

——丹尼尔·马克汉姆（Daniel Markham）

Bedford Technology 公司负责人

序

2003年春，在我公司各个技术团队引入Scrum后不久，我见到了鲍勃大叔。那时我还是个新鲜出炉、心怀疑虑的ScrumMaster。鲍勃教我们使用TDD和一个叫作FitNesse的小工具。我问自己："为什么总要写注定先面临失败的测试用例？测试不该排在编码之后吗？"就像团队中许多其他成员一样，我常常只能挠着头离开。但是，直至现在，鲍勃大叔对编程匠艺的热情于我仍然记忆犹新。他是个直言不讳的人。记得有一天，他在看了我们的缺陷列表后，问我们到底为什么会对并不属于个人的软件系统做出如此糟糕的决定——"这些系统是公司资产，不是你们的个人资产。"他的激情鼓舞了我们。一年半之后，我们实现了80%的自动测试覆盖率，得到了整洁又直观的代码库，客户和团队成员也都满意。之后，我们迅速修正了对"完成"的定义，以之为盾，挡住了潜伏在代码中的小魔怪。本质上，我们学会了如何避免自残。相处日长，我们对鲍勃心生暖意。对我们而言，他如同亲叔父——温暖、坚定、勇敢，一直帮助我们学会自己站起来并做正确的事。有些孩子的"鲍勃大叔"教他们骑单车或钓鱼，而我们这位鲍勃大叔则教我们坚守正直——直至今日，在我的职业生涯中，有能力和愿望，满怀勇气与好奇心地去面对任何环境，仍是鲍勃大叔教会我的最佳课程。

开始从事敏捷教练职业后，我将鲍勃早年教我的那些东西用在工作中，我发现，最好的产品开发团队总能在各种行业、各种客户的各种独特环境中组合不同的最佳实践手段。我还发现，再好的开发工具也需要有与之匹配的人类操作者——那些在不同领域中都能找到这些工具最佳应用方式的团队。当然，我也观察到，开发团队也许达到了很高的单元测试覆盖率，已经能满足指标要求，却发现大部分测试不合格——指标满足，价值不足。最好的团队并不真需要关心指标，他们自有目标、纪律、尊严与责任感，指标自然而然得到满足。《匠艺整洁之道》将这些课程与原则放到具体代码范例与经验讲述中，展示了“为满足交付日期要求而写代码”与“真正搭建未来能用上的系统”之间的区别。

《匠艺整洁之道》提醒我们永不能满足于现状，要*无畏地活着*。这本书就像一位老友，会提醒你什么重要、什么有效、什么无效、什么导致风险、什么降低风险。这些经验历久弥新。你可能会发现自己已经在实践其中的一些技巧，我敢说你会发现另一些新东西，或者至少是你曾因交付压力或其他职业生涯中的压力而放弃了的东西。如果你是开发领域的新手——无论是商业方面的还是技术方面的——那么你将从最优秀的人那里学到东西。即使是最有经验和战斗力的人，也会找到改进自己的方法。也许这本书会帮助你找回激情，重新激起你提升手艺的欲望，或者让你重新投入精力，无惧障碍，追求完美。

软件开发者统治着世界。鲍勃大叔在这里重申了这些“掌握权柄”之人该遵守的职业纪律。他延续了《代码整洁之道》未完的话题。软件开发人员实际上是在编写人类的规则，所以鲍勃大叔提醒我们，必须严守道德准则，有责任知道代码的作用，人们如何使用它们，以及它们会在什么地方出错。软件出错的代价是人的生计——甚至生命。软件影响着我们的思维方式，影响着我们的决定。作为人工智能和预测分析的结果，软件同样影响着社会和人群的行为。因此，我们必须负起责任，以极大的谨慎和同情心行事——人们的健康和福祉取决于此。鲍勃大叔帮助我们面对这种责任，并成为*社会所期望和需要的专业人士*。

在写这篇序的时候，《敏捷宣言》即将迎来它的 20 岁生日[1]。这本书是回归根本的完美机会：它及时而谦逊地提醒我们：程序化世界越来越复杂。为了人类的遗产，也为了我们自己，应该建立和维护职业操守。读读《匠艺整洁之道》吧，让这些原则渗入你的内心，并实践和改进它们，辅导他人。把这本书放在手边书架上，当你带着好奇心和勇气行走于世间时，让这本书成为你的老朋友、你的鲍勃大叔和你的导师吧。

——斯塔西·海格纳·韦斯卡迪（Stacia Heimgartner Viscardi）

CST 和敏捷教练

1 指 2021 年。——编辑注

前　　言

在开始之前，有两个问题需要面对。搞清楚这两个问题，读者才能理解本书所根植的理念。

关于“匠艺”（Craftsmanship）

21 世纪之初的那些年，言辞之争不绝于耳。身在软件行业，我们见证了这些争议。其中，“匠人”（craftsman）一词常被认为太过狭隘。

我思考了很久，与持各种意见的朋友交流。我的结论是，对于本书而言，没有更好的词可用。

我考虑过改用 craftsperson、craftsfolk、crafter 等词，但这些词承担不起 craftsman 一词的历史庄严感。而这种历史庄严感正是本书想传递的重要讯息。

“匠人”让人想到一位技艺高超、成就非凡的行家——善用工具，熟悉行业，为自己的工作而自豪，满怀尊严和专业精神，值得信赖。

你们中的一些人可能会不同意我用这个词，我很理解。我只希望你们无论如何都不要认为这是在试图找到一个非它不可的词，因为这绝不是我本意。

唯一真路

当阅读《匠艺整洁之道》一书时，你可能会感到这是通往工匠精神的唯一真路。对我来说可能是这样，但对你来说可未必。这本书展示了我的路径。当然，你要选择自己的路径。

我们最终会不会需要唯一真路？不知道，也许吧。正如你将读到的那样，对软件职业做出严格定义的难度正在增加。我们也许可以根据所创建的软件的关注重点，采用几种不同的路径。但是，正如你将在下文中读到的那样，要把关键软件和非关键软件区分开来可能并不那么容易。

但我可以肯定一件事。“士师”[1]的日子已一去不返。每名程序员都各自做自己眼中正确的事，已经不够。纪律、标准和对职业操守的要求将会出现。今天摆在我们面前的问题是，让程序员自己来定义这些纪律、标准和职业操守，还是让那些不了解我们的人强加给我们。

本书介绍

本书是为程序员和管程序员的人编写的。但在另一种意义上，本书是为整个人类社会编写的。因为正是我们，这些程序员，无意中发现自己恰好处于这个社会的支点上。

为了自己

如果你已经编程好几年，大概能体会到系统成功部署和运转所带来的满足感。获得这样的成就，作为其中一分子，颇值得骄傲。你为自己能做出这套系统而自豪。

然而，你会为自己做出系统的方式而自豪吗？是为完成了工作而自豪，还是为自己的技艺而自豪？是因为系统得以部署而自豪，还是为你打造系统的方式而自豪？

艰难编程一整天，回到家里，你是会对着镜子里的自己说：“今天干得真棒。”还是只能想到去冲个澡？

当一天结束时，很多程序员会感觉自己很脏。我们觉得自己深陷低水准工作的泥潭。我们

1　源自《旧约·士师记》，是古犹太人对领袖的称呼。

感到，只有牺牲质量才能赶上进度，而且有人在期待我们这样做。我们甚至开始相信，生产力与质量就是相冲突的。

在本书中，我将尽力打破这种思维模式。本书关注如何做好工作。本书将阐述每名程序员都该懂得的纪律与实践手段，遵守这些纪律与掌握相差手段，才能高效工作，并且为自己每天写的代码感到自豪。

为了社会

21 世纪，为了生存，我们的社会开始由无纪律和不受控的技术主导，这是人类历史上首次出现的状况。软件入侵了现代生活的方方面面，从早晨喝咖啡到晚间娱乐，从洗衣到开车。软件让我们既在世界级网络中连接，又在社会和政治层面上分裂。现代世界的生活没有哪一方面不由软件所主导。然而，我们这些构建软件的人不过是乌合之众，对自己所做之事了解甚少。

如果我们这些程序员做得更像样，2020 年艾奥瓦州党内选举结果能否如期得出？两架波音 737 Max 飞机上的 346 位乘客还会罹难吗？骑士资本集团（Knight Capital Group）会在 45 分钟之内损失 4.6 亿美元吗？丰田汽车的意外加速故障会导致 89 人死亡吗？

全世界程序员数量每 5 年翻一番。程序员们几乎没有接受过相关技能教育。他们只是看了看工具，做过几个玩具式的开发项目任务，便被扔进指数级增长的劳动力队伍中，去应付指数级增长的软件需求。每一天，我们称之为软件的那个纸牌屋都在不断深入我们的基础设施、我们的机构、我们的政府，还有我们的生活。每一天，灾难风险都在不断增加。

我说的是什么灾难？不是文明的崩塌，也不是所有软件系统突然解体。摇摇欲坠的纸牌屋并非由软件系统本身构成。我说的是，软件的公众信任基础非常脆弱、岌岌可危。

有太多波音 737 Max 事故，太多丰田汽车意外加速故障，太多加州大众 EPA 丑闻和艾奥瓦州党内选举结果拖延——太多太多臭名昭著的软件失误或恶行。失去信任感、深感愤怒的公众将把目光投向我们的纪律、操守与标准缺失。规条随之而来，那将是我们本不该背负的规条。规条将削弱我们自由探索和延展软件开发工艺的能力，将严厉限制技术发展与经济增长。

本书并不打算阻止人们一头扎进越来越多的软件应用中，也不打算减缓软件生产的速度，因为这种意图注定徒劳无功。社会需要软件，而且无论如何都会得到软件。试图扼杀这种需求，并不能叫停迫在眉睫的公众信任灾难。

相反，本书的目标是让软件开发者和他们的管理者明白纪律的必要性，向他们传授最有效的纪律、标准与职业操守，令他们能够最大限度地生产健壮、高容错和高效的软件。唯有改变我们这些程序员的工作方式，提高纪律、标准和职业操守的水准，才能支撑起纸牌屋，防止它倒塌。

本书结构

本书分为三个部分：纪律、标准、职业操守。

纪律是最基础的一层。这部分关注实用性、技术性和规范性。阅读和理解这个部分，各类程序员都能从中受益。这部分内容配了一些视频[1]，以展示真实的由测试驱动的开发节奏和重构纪律。本书的文字部分也旨在展示这种节奏，但还是视频比较有效。

标准是中间层次。这部分概括了世界对程序员这行的期望。管理者应该好好阅读，从而了解对专业程序员应有的期望。

操守在最高层。这部分阐述了编程职业的道德背景。它以誓言或一套承诺的形式体现，其中包括大量关于历史与哲学的话题。程序员和管理者都应该阅读这部分内容。

给管理者的话

本书包含了对你有益的大量信息。其中也会有你大概不需要理解的大量技术内容。建议你阅读每章的简介部分，当遇到超出所需的技术内容时尽管跳过，直接阅读后续章节。

一定要读第Ⅱ部分“标准”和第Ⅲ部分“操守”。这两部分中的五项纪律都要好好阅读。

1　书中视频可通过扫描本书封底二维码获取。

致　　谢

谢谢勇敢的审阅者们：戴门・波尔（Damon Poole）、埃里克・克里奇劳（Eric Crichlow）、海瑟・坎瑟、蒂姆・奥廷格、杰夫・兰格（Jeff Langr）和斯塔西・韦斯卡迪（Stacia Viscardi）。

感谢朱莉・费弗（Julie Phifer）、克里斯・赞恩（Chris Zahn）、曼卡・麦塔（Menka Mehta）、卡罗尔・莱利尔（Carol Lallier），以及 Pearson 公司所有为本书能顺利出版而殚精竭虑的同人们。

和以往一样，要感谢创意无穷、天才横溢的插画师詹妮弗・孔科（Jennifer Kohnke）。她的作品总令我会心微笑。

当然，还要感谢我深爱的妻子和美好的家庭。

关于作者

1964 年，年仅 12 岁的**罗伯特·C. 马丁（鲍勃大叔）**就已写下他的第一行代码。他自 1970 年起从事程序员职业。他与人合办了 cleancoders.com 网站，为软件开发者提供在线视频培训服务。他还创办了 Uncle Bob 咨询有限公司，为遍布于世界各地的大公司提供软件咨询、培训和技能培养服务。同时，他也供职于芝加哥的软件咨询企业 8th Light，任大匠（Master Craftsman）一职。

马丁先生在多本行业杂志上发表过数十篇文章。他是各种国际性会议和行业活动讲坛上的常客。他也是 cleancoders.com 网站上广受赞誉的多个系列视频的创作者。

马丁先生编著了多本图书，包括：

Designing Object-Oriented C++ Applications Using the Booch Method

Patterns Languages of Program Design 3

More C++ Gems

Extreme Programming in Practice

Agile Software Development: Principles, Patterns, and Practices

UML for Java Programmers

Clean Code

The Clean Coder

Clean Architecture: A Craftsman's Guide to Software Structure and Design

Clean Agile: Back to Basics

作为软件开发行业的领军人物，马丁先生曾任 *C++ Report* 杂志主编达三年之久。他也是敏捷联盟（Agile Alliance）的首任主席。

目　录

1 CRAFTSMANSHIP

（匠艺）

The dream of flying is almost certainly as old as humanity. The ancient Greek myth describing the flight of Daedalus and Icarus dates from around 1550 BCE. In the millennia that followed, a number of brave, if foolish, souls have strapped ungainly contraptions to their bodies and leapt off cliffs and towers to their doom in pursuit of that dream.

Things began to change about five hundred years ago when Leonardo DaVinci drew sketches of machines that, though not truly capable of flight, showed some reasoned thought. It was DaVinci who realized that flight could be possible because air resistance works in both directions. The resistance caused by pushing down on the air creates lift of the same amount. This is the mechanism by which all modern airplanes fly.

DaVinci's ideas were lost until the middle of the eighteenth century. But then began an almost frantic exploration into the possibility of flight. The eighteenth and nineteenth centuries were a time of intense aeronautical research and experimentation. Unpowered prototypes were built, tried, discarded, and improved. The science of aeronautics began to take shape. The forces of lift, drag, thrust, and gravity were identified and understood. Some brave souls made the attempt.

And some crashed and died.

In the closing years of the eighteenth century, and for the half century that followed, Sir George Cayley, the father of modern aerodynamics, built experimental rigs, prototypes, and full-sized models culminating in the first manned flight of a glider.

And some still crashed and died.

Then came the age of steam and the possibility of powered manned flight. Dozens of prototypes and experiments were performed. Scientists and enthusiasts alike joined the gaggle of people exploring the potential of flight. In 1890, Clément Ader flew a twin-engine steam-powered machine for 50 meters.

And some still crashed and died.

But the internal combustion engine was the real game-changer. In all likelihood, the first powered and controlled manned flight took place in 1901 by Gustave Whitehead. But it was the Wright Brothers who, on December 17, 1903, at Kill Devil Hills, North Carolina, conducted the first truly sustained, powered, and controlled manned flight of a heavier-than-air machine.

And some still crashed and died.

But the world changed overnight. Eleven years later, in 1914, biplanes were dogfighting in the air over Europe.

And though many crashed and died under enemy fire, a similar number crashed and died just learning to fly. The principles of flight might have been mastered, but the *technique* of flight was barely understood.

Another two decades, and the truly terrible fighters and bombers of World War II were wreaking havoc over France and Germany. They flew at extreme altitudes. They bristled with guns. They carried devastating destructive power.

During the war, 65,000 American aircraft were lost. But only 23,000 of those were lost in combat. The pilots flew and died in battle. But more often they flew and died when no one was shooting. We still didn't know *how* to fly.

Another decade saw jet-powered craft, the breaking of the sound barrier, and the explosion of commercial airlines and civilian air travel. It was the beginning of the jet age, when people of means (the so-called jet set) could leap from city to city and country to country in a matter of hours.

And the jet airliners tore themselves to shreds and fell out of the sky in terrifying numbers. There was so much we still didn't understand about making and flying aircraft.

That brings us to the 1950s. Boeing 707s would be flying passengers from here to there across the world by the end of the decade. Two more decades would see the first wide-body jumbo jet, the 747.

Aeronautics and air travel settled down to become the safest and most efficient means of travel in the history of the world. It took a long time, and cost many lives, but we had finally learned how to safely build and fly airplanes.[1]

Chesley Sullenberger was born in 1951 in Denison, Texas. He was a child of the jet age. He learned to fly at age sixteen and eventually flew F4 Phantoms for the Air Force. He became a pilot for US Airways in 1980.

On January 15, 2009, just after departure from LaGuardia, his Airbus A320 carrying 155 souls struck a flock of geese and lost both jet engines. Captain Sullenberger, relying on over twenty thousand hours of experience in the air, guided his disabled craft to a "water landing" in the Hudson River and, through sheer indomitable skill, saved every one of those 155 souls. Captain Sullenberger excelled at his craft. Captain Sullenberger was a craftsman.

The dream of fast, reliable computation and data management is almost certainly as old as humanity. Counting on fingers, sticks, and beads dates back thousands of years. People were building and using abacuses over four thousand years ago. Mechanical devices were used to predict the movement of stars and planets some two thousand years ago. Slide rules were invented about four hundred years ago.

In the early nineteenth century, Charles Babbage started building crank-powered calculating machines. These were true digital computers with memory and arithmetic processing. But they were difficult to build with the metalworking technology of the day, and though he built a few prototypes, they were not a commercial success.

1. The 737 Max notwithstanding.

In the mid-1800s, Babbage attempted to build a much more powerful machine. It would have been steam powered and capable of executing true programs. He dubbed it the *Analytical Engine*.

Lord Byron's daughter, Ada—the Countess of Lovelace—translated the notes of a lecture given by Babbage and discerned a fact that had apparently not occurred to anyone else at the time: *the numbers in a computer need not represent numbers at all but can represent things in the real world*. For that insight, she is often called the world's first true programmer.

Problems of precise metalworking continued to frustrate Babbage, and in the end, his project failed. No further progress was made on digital computers throughout the rest of the nineteenth and early twentieth centuries. During that time, however, mechanical *analog* computers reached their heyday.

In 1936, Alan Turing showed that there is no general way to prove that a given Diophantine[2] equation has solutions. He constructed this proof by imagining a simple, if infinite, digital computer and then proving that there were numbers that this computer could not calculate. As a consequence of this proof, he invented finite state machines, machine language, symbolic language, macros, and primitive subroutines. He invented, what we would call today, software.

At almost exactly the same time, Alonzo Church constructed a completely different proof for the same problem and consequently developed the lambda calculus—the core concept of functional programming.

In 1941, Konrad Zuse built the first electromechanical programmable digital computer, the Z3. It consisted of more than 2,000 relays and operated at a clock rate of 5 to 10Hz. The machine used binary arithmetic organized into 22-bit words.

During World War II, Turing was recruited to help the "boffins" at Bletchley Park decrypt the German Enigma codes. The Enigma machine was a simple

2. Equations of integers.

digital computer that randomized the characters of textual messages that were typically broadcast using radio telegraphs. Turing aided in the construction of an electromechanical digital search engine to find the keys to those codes.

After the war, Turing was instrumental in building and programming one of the world's first electronic vacuum tube computers—the Automatic Computing Engine, or ACE. The original prototype used 1,000 vacuum tubes and manipulated binary numbers at a speed of a million bits per second.

In 1947, after writing some programs for this machine and researching its capabilities, Turing gave a lecture in which he made these prescient statements:

> *We shall need a great number of mathematicians of ability [to put the problems] into a form for computation.*
>
> *One of our difficulties will be the maintenance of an appropriate discipline, so that we do not lose track of what we are doing.*

And the world changed overnight.

Within a few years, core memory had been developed. The possibility of having hundreds of thousands, if not millions, of bits of memory accessible within microseconds became a reality. At the same time, mass production of vacuum tubes made computers cheaper and more reliable. Limited mass production was becoming a reality. By 1960, IBM had sold 140 model 70x computers. These were huge vacuum tube machines worth millions of dollars.

Turing had programmed his machine in binary, but everyone understood that was impractical. In 1949, Grace Hopper had coined the word *compiler* and by 1952 had created the first one: A-0. In late 1953, John Bachus submitted the first FORTRAN specification. ALGOL and LISP followed by 1958.

The first working transistor was created by John Bardeen, Walter Brattain, and William Shockley in 1947. They made their way into computers in 1953. Replacing vacuum tubes with transistors changed the game entirely. Computers became smaller, faster, cheaper, and much more reliable.

By 1965, IBM had produced 10,000 model 1401 computers. They rented for $2,500 per month. This was well within the reach of midsized businesses. Those businesses needed programmers, and so the demand for programmers began to accelerate.

Who was programming all these machines? There were no university courses. Nobody went to school to learn to program in 1965. These programmers were drawn from business. They were mature folks who had worked in their businesses for some time. They were in their 30s, 40s, and 50s.

By 1966, IBM was producing 1,000 model 360 computers every month. Businesses could not get enough. These machines had memory sizes that reached 64kB and more. They could execute hundreds of thousands of instructions per second.

That same year, working on a Univac 1107 at the Norwegian Computer Center, Ole-Johan Dahl and Kristen Nygard invented Simula 67, an extension of ALGOL. It was the first object-oriented language.

Alan Turing's lecture was only two decades in the past!

Two years later, in March 1968, Edsger Dijkstra wrote his famous letter to the *Communications of the ACM* (*CACM*). The editor gave that letter the title "Go To Statement Considered Harmful."[3] Structured programming was born.

In 1972, at Bell Labs in New Jersey, Ken Thompson and Dennis Ritchie were between projects. They begged time on a PDP 7 from a different project team and invented UNIX and C.

Now the pace picked up to near breakneck speeds. I'm going to give you a few key dates. For each one, ask yourself how many computers are in the

3. Edsger W. Dijkstra, "Go To Statement Considered Harmful," *Communications of the ACM* 11, no. 3 (1968).

world? How many programmers are in the world? And where did those programmers come from?

1970—Digital Equipment Corporation had produced 50,000 PDP-8 computers since 1965.

1970—Winston Royce wrote the "waterfall" paper, "Managing the Development of Large Software Systems."

1971—Intel released the 4004 single-chip microcomputer.

1974—Intel released the 8080 single-chip microcomputer.

1977—Apple released the Apple II.

1979—Motorola released the 68000, a 16-bit single-chip microcomputer.

1980—Bjarne Stroustrup invented *C with Classe*s (a preprocessor that makes C look like Simula).

1980—Alan Kay invented Smalltalk.

1981—IBM released the IBM PC.

1983—Apple released the 128K Macintosh.

1983—Stroustrup renamed C with Classes to C++.

1985—The US Department of Defense adopted waterfall as the official software process (DOD-STD-2167A).

1986—Stroustrup published *The C++ Programming Language* (Addison-Wesley).

1991—Grady Booch published *Object-Oriented Design with Applications* (Benjamin/Cummings).

1991—James Gosling invented Java (called *Oak* at the time).

1991—Guido Van Rossum released Python.

1995—*Design Patterns: Elements of Reusable Object-Oriented Software* (Addison-Wesley) was written by Erich Gamma, Richard Helm, John Vlissides, and Ralph Johnson.

1995—Yukihiro Matsumoto released Ruby.

1995—Brendan Eich created JavaScript.

1996—Sun Microsystems released Java.

1999—Microsoft invented C#/.NET (then called *Cool*).

2000—Y2K! The Millennium Bug.

2001—The Agile Manifesto was written.

Between 1970 and 2000, the clock rates of computers increased by three orders of magnitude. Density increased by four orders of magnitude. Disk space increased by six or seven orders of magnitude. RAM capacity increased by six or seven orders of magnitude. Costs had fallen from dollars per bit to dollars per gigabit. The change in the hardware is hard to visualize, but just summing up all the orders of magnitude I mentioned leads us to about a thirty orders of magnitude increase in capability.

And all this in just over fifty years since Alan Turing's lecture.

How many programmers are there now? How many lines of code have been written? How good is all that code?

Compare this timeline with the aeronautical timeline. Do you see the similarity? Do you see the gradual increase in theory, the rush and failure of the enthusiasts, the gradual increase in competence? The decades of not knowing what we were doing?

And now, with our society depending, for its very existence, on our skills, do we have the Sullenbergers whom our society needs? Have we groomed the programmers who understand their craft as deeply as today's airline pilots understand theirs? Do we have the craftsmen whom we shall certainly require?

Craftsmanship is the state of knowing how to do something well and is the outcome of good tutelage and lots of experience. Until recently, the software industry had far too little of either. Programmers tended not to remain programmers for long, because they viewed programming as a steppingstone into management. This meant that there were few programmers who acquired enough experience to teach the craft to others. To make matters worse, the number of new programmers entering the field doubles every five years or so, keeping the ratio of experienced programmers far too low.

The result has been that most programmers never learn the disciplines, standards, and ethics that could define their craft. During their relatively brief career of writing code, they remain unapprenticed novices. And this, of course, means that much of the code produced by those inexperienced programmers is substandard, ill structured, insecure, buggy, and generally a mess.

In this book, I describe the standards, disciplines, and ethics that I believe every programmer should know and follow in order to gradually acquire the knowledge and skill that their craft truly requires.

I THE DISCIPLINES

（纪律）

What is a discipline? A discipline is a set of rules that are composed of two parts: the essential part and the arbitrary part. The essential part is what gives the discipline its power; it is the reason that the discipline exists. The arbitrary part is what gives the discipline its form and substance. The discipline cannot exist without the arbitrary part.

For example, surgeons wash their hands before surgery. If you were to watch, you would see that the handwashing has a very particular form to it. The surgeon does not wash hands by simply soaping them under running water, as you and I might do. Rather, the surgeon follows a ritualized discipline of handwashing. One such routine I have seen is, in part, as follows:

- Use the proper soap.
- Use the appropriate brush.
- For each finger, use
  - Ten strokes across the top.
  - Ten strokes across the left side.
  - Ten strokes across the underside.
  - Ten strokes across the right side.
  - Ten strokes across the nail.
- And so on.

The essential part of the discipline should be obvious. The surgeon's hands must be made very clean. But did you notice the arbitrary part? Why ten strokes instead of eight or twelve? Why divide the finger into five sections? Why not three or seven sections?

That's all arbitrary. There is no real reason for those numbers other than that they were deemed to be sufficient.

In this book, we study five disciplines of software craftsmanship. Some of these disciplines are five decades old. Some are just two decades old. But all have shown their usefulness over those decades. Without them, the very notion of software-as-a-craft would be virtually unthinkable.

Each of these disciplines has its own essential and arbitrary elements. As you read, you may find your mind objecting to one or more of the disciplines. If this happens, be aware of whether the objection is about the essential elements of the disciplines or just the arbitrary elements. Don't allow yourself to be misdirected by the arbitrary elements. Keep your focus on the essential elements. Once you have internalized the essence of each discipline, the arbitrary form will be likely to diminish in importance.

For example, in 1861, Ignaz Semmelweis published his findings for applying the discipline of handwashing for doctors. The results of his research were astounding. He was able to show that when doctors thoroughly washed their hands in chlorine bleach before examining pregnant women, the death rates of those women from subsequent sepsis dropped from one in ten to virtually zero.

But the doctors of the day did not separate the essence from the arbitrary when reviewing Semmelweis's proposed discipline. The chlorine bleach was the arbitrary part. The washing was the essence. They were repelled by the inconvenience of washing with bleach, and so they rejected the evidence of the essential nature of handwashing.

It was many decades before doctors started actually washing their hands.

Extreme Programming（极限编程）

In 1970, Winston Royce published the paper that drove the waterfall development process into the mainstream. It took almost 30 years to undo that mistake.

By 1995, software experts started considering a different, more incremental approach. Processes such as Scrum, feature-driven development (FDD), dynamic systems development method (DSDM), and the Crystal methodologies were presented. But little changed in the industry at large.

Then, in 1999, Kent Beck published the book *Extreme Programming Explained* (Addison-Wesley). Extreme Programming (XP) built upon the ideas in those previous processes but added something new. XP added *engineering practices*.

Enthusiasm grew exponentially for XP between 1999 and 2001. It was this enthusiasm that spawned and drove the Agile revolution. To this day, XP remains the best defined and most complete of all the Agile methods. The engineering practices at its core are the focus of this section on disciplines.

The Circle of Life（生命之环）

In Figure I.1, you see Ron Jeffries' *Circle of Life,* which shows the practices of XP. The disciplines that we cover in this book are the four in the center and the one on the far left.

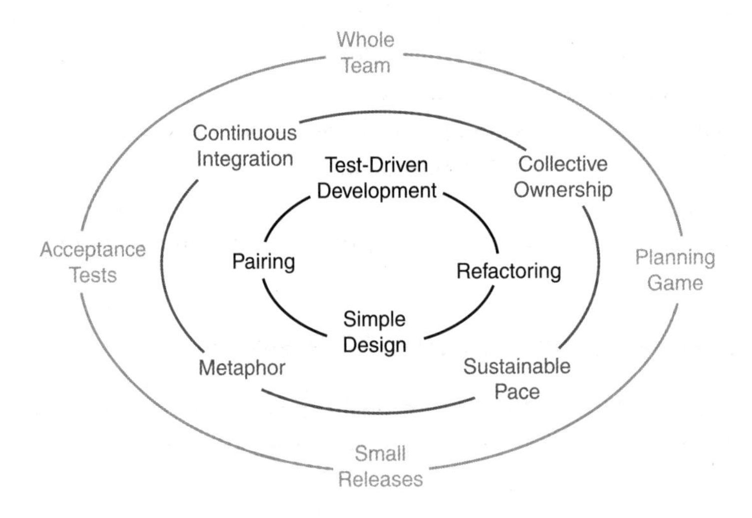

Figure I.1 The Circle of Life: The practices of XP

The four in the center are the engineering practices of XP: test-driven development (TDD), refactoring, simple design, and pairing (which we shall call *collaborative programming*. The practice at the far left, acceptance tests, is the most technical and engineering focused of the business practices of XP.

These five practices are among the foundational disciplines of software craftsmanship.

Test-Driven Development（测试驱动开发）

TDD is the lynchpin discipline. Without it, the other disciplines are either impossible or impotent. For that reason, the two upcoming sections describing TDD represent nearly half the pages of this book and are intensely technical. This organization may seem unbalanced. Indeed, it feels that way to me too, and I struggled with what to do about that. My conclusion, however, is that the imbalance is a reaction to the corresponding imbalance within our industry. Too few programmers know this discipline well.

TDD is the discipline that governs the way a programmer works on a second-by-second basis. It is neither an up-front discipline nor an after-the-fact discipline. TDD is in-process and in-your-face. There is no way to do partial TDD; it is an all-or-nothing discipline.

The essence of the TDD discipline is very simple. Small cycles and tests come first. Tests come first in everything. Tests are written first. Tests are cleaned up first. In all activities, tests come first. And all activities are broken down into the tiniest of cycles.

Cycle times are measured in seconds, not minutes. They are measured in characters, not lines. The feedback loop is closed almost literally as soon as it is opened.

The goal of TDD is to create a test suite that you would trust with your life. If the test suite passes, you should feel safe to deploy the code.

Of all the disciplines, TDD is the most onerous and the most complex. It is onerous because it dominates everything. It is the first and last thing you think about. It is the constraint that lays thick over everything you do. It is the governor that keeps the pace steady regardless of pressure and environmental stresses.

TDD is complex because code is complex. For each shape or form of code, there is a corresponding shape or form of TDD. TDD is complex because the tests must be designed to fit the code without being coupled to the code and must cover almost everything yet still execute in seconds. TDD is an elaborate and complex skill that is very hard won but immeasurably rewarding.

Refactoring（重构）

Refactoring is the discipline that allows us to write clean code. Refactoring is difficult, if not impossible, without TDD.[1] Therefore, writing clean code is just as difficult or impossible without TDD.

Refactoring is the discipline by which we manipulate poorly structured code into code with a better structure—*without affecting behavior*. That last part is critical. By guaranteeing that the behavior of the code is unaffected, the improvements in structure are guaranteed to be *safe*.

The reason we don't clean up code—the reason that software systems rot over time—is that we are afraid that cleaning the code will break the behavior. But if we have a way to clean up code that we know is safe, then *we will* clean up the code, and our systems will not rot.

How do we guarantee that our improvements do not affect behavior? We have the tests from TDD.

Refactoring is also a complex discipline because there are many ways to create poorly structured code. Thus, there are many strategies for cleaning up that code. Moreover, each of those strategies must fit frictionlessly and concurrently into the TDD test-first cycle. Indeed, these two disciplines are so deeply intertwined that they are virtually inseparable. It is almost impossible to refactor without TDD, and it is virtually impossible to practice TDD without practicing refactoring.

1. There may be other disciplines that could support refactoring as well as TDD. Kent Beck's test && commit || revert is a possibility. At the time of this writing, however, it has not enjoyed a high degree of adoption and remains more of an academic curiosity.

Simple Design（简单设计）

Life on Earth could be described in terms of layers. At the top is ecology, the study of systems of living things. Below that is physiology, the study life's internal mechanisms. The next layer down might be microbiology, the study of cells, nucleic acids, proteins, and other macromolecular systems. Those are in turn described by the science of chemistry, which in turn is described by quantum mechanics.

Extending that analogy to programming, if TDD is the quantum mechanics of programming, then refactoring is the chemistry and simple design is the microbiology. Continuing that analogy, SOLID principles, object-oriented design, and functional programming are the physiology, and architecture is the ecology of programming.

Simple design is almost impossible without refactoring. Indeed, it is the end goal of refactoring, and refactoring is the only practical means to achieve that goal. That goal is the production of simple atomic granules of design that fit well into the larger structures of programs, systems, and applications.

Simple design is not a complex discipline. It is driven by four very simple rules. However, unlike TDD and refactoring, simple design is an imprecise discipline. It relies on judgment and experience. Done well, it is the first indication that separates an apprentice who knows the rules from a journeyman who understands the principles. It is the beginning of what Michael Feathers has called *design sense*.

Collaborative Programming（协同编程）

Collaborative programming is the discipline and the art of working together in a software team. It includes subdisciplines such as pair programming, mob programming, code reviews, and brainstorms. Collaborative programming involves everyone on the team, programmers and nonprogrammers alike. It is the primary means by which we share knowledge, assure consistency, and gel the team into a functioning whole.

Of all the disciplines, collaborative programming is the least technical and the least prescriptive. Nevertheless, it may be the most important of the five disciplines, because the building of an effective team is both a rare and precious thing.

Acceptance Tests（验收测试）

Acceptance testing is the discipline that ties the software development team to the business. The business purpose is the specification of the desired behaviors of the system. Those behaviors are encoded into tests. If those tests pass, the system behaves as specified.

The tests must be readable and writeable by representatives of the business. It is by writing and reading these tests and seeing them pass that the business knows what the software does and that it does what the business needs it to do.

2 TEST-DRIVEN DEVELOPMENT

（测试驱动开发）

Our discussion of test-driven development (TDD) spans two chapters. We first cover the basics of TDD in a very technical and detailed manner. In this chapter, you will learn about the discipline in a step-by-step fashion. The chapter provides a great deal of code to read and several videos to watch as well.

In Chapter 3, "Advanced TDD," we cover many of the traps and conundrums that novice TDDers face, such as databases and graphical user interfaces. We also explore the design principles that drive good test design and the design patterns of testing. Finally, we investigate some interesting and profound theoretical possibilities.

OVERVIEW（概述）

Zero. It's an important number. It's the number of balance. When the two sides of a scale are in balance, the pointer on the scale reads zero. A neutral atom, with equal numbers of electrons and protons, has a charge of zero. The sum of forces on a bridge balances to zero. Zero is the number of balance.

Did you ever wonder why the amount of money in your checking account is called its balance? That's because the balance in your account is the sum of all the transactions that have either deposited or withdrawn money from that account. But transactions always have two sides because transactions move money *between* accounts.

The *near* side of a transaction affects your account. The *far* side affects some other account. Every transaction whose near side deposits money into your account has a far side that withdraws that amount from some other account. Every time you write a check, the near side of the transaction withdraws money from your account, and the far side deposits that money into some other account. So, the balance in *your* account is the sum of the near sides of the transactions. The sum of the far sides should be equal and opposite to the balance of your account. The sum of all the near and far sides should be zero.

Two thousand years ago, Gaius Plinius Secundus, known as Pliny the Elder, realized this law of accounting and invented the discipline of double-entry bookkeeping. Over the centuries, this discipline was refined by the bankers in Cairo and then by the merchants of Venice. In 1494, Luca Pacioli, a Franciscan friar and friend of Leonardo DaVinci, wrote the first definitive description of the discipline. It was published in book form on the newly invented printing press, and the technique spread.

In 1772, as the industrial revolution gained momentum, Josiah Wedgwood was struggling with success. He was the founder of a pottery factory, and his product was in such high demand that he was nearly bankrupting himself trying to meet that demand. He adopted double-entry bookkeeping and was thereby able to see how money was flowing in and out of his business with a resolution that had previously escaped him. And by tuning those flows, he staved off the looming bankruptcy and built a business that exists to this day.

Wedgwood was not alone. Industrialization drove the vast growth of the economies of Europe and America. In order to manage all the money flows resulting from that growth, increasing numbers of firms adopted the discipline.

In 1795, Johann Wolfgang von Goethe wrote the following in *Wilhelm Meister's Apprenticeship*. Pay close attention, for we will return to this quote soon.

> *"Away with it, to the fire with it!" cried Werner. "The invention does not deserve the smallest praise: that affair has plagued me enough already, and drawn upon yourself your father's wrath. The verses may be altogether beautiful; but the meaning of them is fundamentally false. I still recollect your Commerce personified; a shrivelled, wretched-looking sibyl she was. I suppose you picked up the image of her from some miserable huckster's shop. At that time, you had no true idea at all of trade; whilst I could not think of any man whose spirit was, or needed to be, more enlarged than the spirit of a genuine merchant. What a thing it is to see the order which prevails throughout his business! By means of this he can at any time survey the general whole, without needing to perplex himself in the details. What advantages does he derive from the system of book-keeping by double entry! It is among the finest inventions of the human mind; every prudent master of a house should introduce it into his economy."*

Today, double-entry bookkeeping carries the force of law in almost every country on the planet. To a large degree, the discipline *defines* the accounting profession.

But let's return to Goethe's quote. Note the words that Goethe used to describe the means of "Commerce" that he so detested:

> *A shrivelled, wretched-looking sibyl she was. I suppose you picked up the image of her from some miserable huckster's shop.*

Have you seen any code that matches that description? I'm sure you have. So have I. Indeed, if you are like me, then you have seen far, far too much of it. If you are like me, you have *written* far, far too much of it.

Now, one last look at Goethe's words:

> *What a thing it is to see the order which prevails throughout his business! By means of this he can at any time survey the general whole, without needing to perplex himself in the details.*

It is significant that Goethe ascribes this powerful benefit to the simple discipline of double-entry bookkeeping.

SOFTWARE（软件）

The maintenance of proper accounts is utterly essential for running a modern business, and the discipline of double-entry bookkeeping is essential for the maintenance of proper accounts. But is the proper maintenance of software any less essential to the running of a business? By no means! In the twenty-first century, software is at the heart of every business.

What, then, can software developers use as a discipline that gives them the control and vision over their software that double-entry bookkeeping gives to accountants and managers? Perhaps you think that software and accounting are such different concepts that no correspondence is required or even possible. I beg to differ.

Consider that accounting is something of a mage's art. Those of us not versed in its rituals and arcanities understand but little of the depth of the accounting profession. And what is the work product of that profession? It is a set of documents that are organized in a complex and, for the layperson, bewildering fashion. Upon those documents is strewn a set of symbols that few but the accountants themselves can truly understand. And yet if even one of those symbols were to be in error, terrible consequences could ensue. Businesses could founder and executives could be jailed.

Now consider how similar accounting is to software development. Software is a mage's art indeed. Those not versed in the rituals and arcanities of software development have no true idea of what goes on under the surface. And the product? Again, a set of documents: the source code—documents organized in a deeply complex and bewildering manner, littered with symbols that only the programmers themselves can divine. And if even one of those symbols is in error, terrible consequences may ensue.

The two professions are deeply similar. They both concern themselves with the intense and fastidious management of intricate detail. They both require significant training and experience to do well. They both are engaged in the production of complex documents whose accuracy, at the level of individual symbols, is critical.

Accountants and programmers may not want to admit it, but they are of a kind. And the discipline of the older profession should be well observed by the younger.

As you will see in what follows, TDD *is* double-entry bookkeeping. It is the same discipline, executed for the same purpose, and delivering the same results. Everything is said twice, in complementary accounts that must be kept in balance by keeping the tests passing.

The Three Laws of TDD（TDD三法则）

Before we get to the three laws, we have some preliminaries to cover.

The essence of TDD entails the discipline to do the following:

1. Create a test suite that enables refactoring and is trusted to the extent that passage implies deployability. That is, if the test suite passes, the system can be deployed.
2. Create production code that is decoupled enough to be testable and refactorable.
3. Create an extremely short-cycle feedback loop that maintains the task of writing programs with a stable rhythm and productivity.
4. Create tests and production code that are sufficiently decoupled from each other so as to allow convenient maintenance of both, without the impediment of replicating changes between the two.

The discipline of TDD is embodied within three entirely arbitrary laws. The proof that these laws are arbitrary is that the essence can be achieved by very different means. In particular, Kent Beck's test && commit || revert (TCR) discipline. Although TCR is entirely different from TDD, it achieves precisely the same essential goals.

The three laws of TDD are the basic foundation of the discipline. Following them is very hard, especially at first. Following them also requires some skill and knowledge that is hard to come by. If you try to follow these laws without that skill and knowledge, you will almost certainly become frustrated and abandon the discipline. We address that skill and knowledge in subsequent chapters. For the moment, be warned. Following these laws without proper preparation will be very difficult.

The First Law

Write no production code until you have first written a test that fails due to the lack of that production code.

If you are a programmer of any years' experience, this law may seem foolish. You might wonder what test you are supposed to write if there's no code to test. This question comes from the common expectation that tests are written *after* code. But if you think about it, you'll realize that if you can write the production code, you can also write the code that tests the production code. It may seem out of order, but there's no lack of information preventing you from writing the test first.

The Second Law

> *Write no more of a test than is sufficient to fail or fail to compile. Resolve the failure by writing some production code.*

Again, if you are an experienced programmer, then you likely realize that the very first line of the test will fail to compile because that first line will be written to interact with code that does not yet exist. And that means, of course, that you will not be able to write more than one line of a test before having to switch over to writing production code.

The Third Law

> *Write no more production code than will resolve the currently failing test. Once the test passes, write more test code.*

And now the cycle is complete. It should be obvious to you that these three laws lock you into a cycle that is just a few seconds long. It looks like this:

- You write a line of test code, but it doesn't compile (of course).
- You write a line of production code that makes the test compile.
- You write another line of test code that doesn't compile.
- You write another line or two of production code that makes the test compile.
- You write another line or two of test code that compiles but fails an assertion.
- You write another line or two of production code that passes the assertion.

And this is going to be your life from now on.

Once again, the experienced programmer will likely consider this to be absurd. The three laws lock you into a cycle that is just a few seconds long.

Each time around that cycle, you are switching between test code and production code. You'll never be able to just write an `if` statement or a `while` loop. You'll never be able to just write a function. You will be forever trapped in this tiny little loop of switching contexts between test code and production code.

You may think that this will be tedious, boring, and slow. You might think that it will impede your progress and interrupt your chain of thought. You might even think that it's just plain silly. You may think that this approach will lead you to produce spaghetti code or code with little or no design—a haphazard conglomeration of tests and the code that makes those tests pass.

Hold all those thoughts and consider what follows.

Losing the Debug-foo

I want you to imagine a room full of people following these three laws—a team of developers all working toward the deployment of a major system. Pick any one of those programmers you like, at any time you like. Everything that programmer is working on executed and passed all its tests within the last minute or so. And this is always true. It doesn't matter who you pick. It doesn't matter when you pick them. Everything worked a minute or so ago.

What would your life be like if everything worked a minute or so ago? How much debugging do you think you would do? The fact is that there's not likely much to debug if everything worked a minute or so ago.

Are you good at the debugger? Do you have the debug-foo in your fingers? Do you have all the hot keys primed and ready to go? Is it second nature for you to efficiently set breakpoints and watchpoints and to dive headlong into a deep debugging session?

This is not a skill to be desired!

You don't want to be good at the debugger. The only way you get good at the debugger is by spending a lot of time debugging. And I don't want you spending a lot of time debugging. You shouldn't want that either. I want you spending as much time as possible writing code that works and as little time as possible fixing code that doesn't.

I want your use of the debugger to be so infrequent that you forget the hot keys and lose the debug-foo in your fingers. I want you puzzling over the obscure step-into and step-over icons. I want you to be so unpracticed at the debugger that the debugger feels awkward and slow. And you should want that too. The more comfortable you feel with a debugger, the more you know you are doing something wrong.

Now, I can't promise you that these three laws will eliminate the need for the debugger. You will still have to debug from time to time. This is still software, and it's still hard. But the frequency and duration of your debugging sessions will undergo a drastic decline. You will spend far more time writing code that works and far less time fixing code that doesn't.

Documentation

If you've ever integrated a third-party package, you know that included in the bundle of software you receive is a PDF written by a tech writer. This document purports to describe how to integrate the third-party package. At the end of this document is almost always an ugly appendix that contains all the *code examples* for integrating the package.

Of course, that appendix is the first place you look. You don't want to read what a tech writer wrote *about* the code; you want to read the code. And that code will tell you much more than the words written by the tech writer. If you are lucky, you might even be able to use copy/paste to move the code into your application where you can fiddle it into working.

When you follow the three laws, you are writing the *code examples* for the whole system. Those tests you are writing explain every little detail about how the system works. If you want to know how to create a certain business object, there are tests that show you how to create it every way that it can be created. If you want to know how to call a certain API function, there are tests that demonstrate that API function and all its potential error conditions and exceptions. There are tests in the test suite that will tell you anything you want to know about the details of the system.

Those tests are documents that describe the entire system at its lowest level. These documents are written in a language you intimately understand. They are utterly unambiguous. They are so formal that they execute. And they cannot get out of sync with the system.

As documents go, they are almost perfect.

I don't want to oversell this. The tests are not particularly good at describing the motivation for a system. They are not high-level documents. But at the lowest level, they are better than any other kind of document that could be written. They are code. And code is something you know will tell you the truth.

You might be concerned that the tests will be as hard to understand as the system as a whole. But this is not the case. Each test is a small snippet of code that is focused on one very narrow part of the system as a whole. The tests do not form a system by themselves. The tests do not know about each other, and so there is no rat's nest of dependency in the tests. Each test stands alone. Each test is understandable on its own. Each test tells you exactly what you need to understand within a very narrow part of the system.

Again, I don't want to oversell this point. It is possible to write opaque and complex tests that are hard to read and understand, but it is not necessary. Indeed, it is one of the goals of this book to teach you how to write tests that are clear and clean documents that describe the underlying system.

Holes in the Design

Have you ever written tests after the fact? Most of us have. Writing tests after writing code is the most common way that tests are written. But it's not a lot of fun, is it?

It's not fun because by the time we start writing after-the-fact tests, we already know the system works. We've tested it manually. We are only writing the tests out of some sense of obligation or guilt or, perhaps, because our management has mandated some level of test coverage. So, we begrudgingly bend into the grind of writing one test after another, knowing that each test we write will pass. Boring, boring, boring.

Inevitably, we come to the test that's hard to write. It is hard to write because we did not design the code to be testable; we were focused instead on making it work. Now, in order to test the code, we're going to have to change the design.

But that's a pain. It's going to take a lot of time. It might break something else. And we already know the code works because we tested it manually. Consequently, we walk away from that test, leaving a hole in the test suite. Don't tell me you've never done this. You know you have.

You also know that if you've left a hole in the test suite, everybody else on the team has too, so you know that the test suite is full of holes.

The number of holes in the test suite can be determined by measuring the volume and duration of the laughter of the programmers when the test suite passes. If the programmers laugh a lot, then the test suite has a lot of holes in it.

A test suite that inspires laughter when it passes is not a particularly useful test suite. It may tell you when certain things break, but there is no decision you can make when it passes. When it passes, all you know is that some stuff works.

A good test suite has no holes. A good test suite allows you to make a decision when it passes. That decision is to *deploy*.

If the test suite passes, you should feel confident in recommending that the system be deployed. If your test suite doesn't inspire that level of confidence, of what use is it?

Fun

When you follow the three laws, something very different happens. First of all, it's fun. One more time, I don't want to oversell this. TDD is not as much fun as winning the jackpot in Vegas. It's not as much fun as going to a party or even playing Chutes and Ladders with your four-year-old. Indeed, *fun* might not be the perfect word to use.

Do you remember when you got your very first program to work? Remember that feeling? Perhaps it was in a local department store that had a TRS-80 or a Commodore 64. Perhaps you wrote a silly little infinite loop that printed your name on the screen forever and ever. Perhaps you walked away from that screen with a little smile on your face, knowing that you were the master of the universe and that all computers would bow down to you forever.

A tiny echo of that feeling is what you get every time you go around the TDD loop. Every test that fails just the way you expected it to fail makes you nod and smile just a little bit. Every time you write the code that makes that failing test pass, you remember that once you were master of the universe and that you still have *the power*.

Every time around the TDD loop, there's a tiny little shot of endorphins released into your reptile brain, making you feel just a little more competent and confident and ready to meet the next challenge. And though that feeling is small, it is nonetheless kinda fun.

Design

But never mind the fun. Something much more important happens when you write the tests first. It turns out that you cannot write code that's hard to test if you write the tests first. The act of writing the test first forces you to design the code to be easy to test. There's no escape from this. If you follow the three laws, your code will be easy to test.

What makes code hard to test? Coupling and dependencies. Code that is easy to test does not have those couplings and dependencies. Code that is easy to test is decoupled!

Following the three laws forces you to write decoupled code. Again, there is no escape from this. If you write the tests first, the code that passes those tests will be decoupled in ways that you'd never have imagined.

And that's a very good thing.

The Pretty Little Bow on Top

It turns out that applying the three laws of TDD has the following set of benefits:

- You will spend more time writing code that works and less time debugging code that doesn't.
- You will produce a set of nearly perfect low-level documentation.
- It is fun—or at least motivating.
- You will produce a test suite that will give you the confidence to deploy.
- You will create less-coupled designs.

These reasons might convince you that TDD is a good thing. They might be enough to get you to ignore your initial reaction, even repulsion. Maybe.

But there is a far more overriding reason why the discipline of TDD is important.

Fear

Programming is hard. It may be the most difficult thing that humans have attempted to master. Our civilization now depends upon hundreds of thousands of interconnected software applications, each of which involves hundreds of thousands if not tens of millions of lines of code. There is no other apparatus constructed by humans that has so many moving parts.

Each of those applications is supported by teams of developers who are scared to death of change. This is ironic because the whole reason software exists is to allow us to easily change the behavior of our machines.

But software developers know that every change introduces the risk of breakage and that breakage can be devilishly hard to detect and repair.

Imagine that you are looking at your screen and you see some nasty tangled code there. You probably don't have to work very hard to conjure that image because, for most of us, this is an everyday experience.

Now let's say that as you glance at that code, for one very brief moment, the thought occurs to you that you ought to clean it up a bit. But your very next thought slams down like Thor's hammer: "I'M NOT TOUCHING IT!" Because you know that if you touch it, you will break it; and if you break it, it becomes *yours forever*.

This is a fear reaction. You fear the code you maintain. You fear the consequences of breaking it.

The result of this fear is that the code must rot. No one will clean it. No one will improve it. When forced to make changes, those changes will be made in the manner that is safest for the programmer, not best for the system. The design will degrade, and the code will rot, and the productivity of the team will decline, and that decline will continue until productivity is near zero.

Ask yourself if you have ever been significantly slowed down by the bad code in your system. Of course you have. And now you know why that bad code exists. It exists because nobody has the courage to do the one thing that could improve it. No one dares risk cleaning it.

Courage

But what if you had a suite of tests that you trusted so much that you were confident in recommending deployment every time that suite of tests passed? And what if that suite of tests executed in seconds? How much would you then fear to engage in a gentle cleaning of the system?

Imagine that code on your screen again. Imagine the stray thought that you might clean it up a little. What would stop you? You have the tests. Those tests will tell you the instant you break something.

With that suite of tests, you can safely clean the code. With that suite of tests, you can *safely* clean the code. *With that suite of tests, you can safely clean the code.*

No, that wasn't a typo. I wanted to drive the point home very, very hard. With that suite of tests, you can safely clean the code!

And if you can safely clean the code, you *will* clean the code. And so will everyone else on the team. Because nobody likes a mess.

The Boy Scout Rule

If you have that suite of tests that you trust with your professional life, then you can safely follow this simple guideline:

> *Check the code in cleaner than you checked it out.*

Imagine if everyone did that. Before checking the code in, they made one small act of kindness to the code. They cleaned up one little bit.

Imagine if every check-in made the code cleaner. Imagine that nobody ever checked the code in worse than it was but always better than it was.

What would it be like to maintain such a system? What would happen to estimates and schedules if the system got cleaner and cleaner with time? How long would your bug lists be? Would you need an automated database to maintain those bug lists?

That's the Reason

Keeping the code clean. Continuously cleaning the code. That's why we practice TDD. We practice TDD so that we can be proud of the work we do.

So that we can look at the code and know it is clean. So that we know that every time we touch that code, it gets better than it was before. And so that we go home at night and look in the mirror and smile, knowing we did a good job today.

The Fourth Law（第四法则）

I will have much more to say about refactoring in later chapters. For now, I want to assert that refactoring is the fourth law of TDD.

From the first three laws, it is easy to see that the TDD cycle involves writing a very small amount of test code that fails, and then writing a very small amount of production code that passes the failing test. We could imagine a traffic light that alternates between red and green every few seconds.

But if we were to allow that cycle to continue in that form, then the test code and the production code would rapidly degrade. Why? Because humans are not good at doing two things at once. If we focus on writing a failing test, it's not likely to be a well-written test. If we focus on writing production code that passes the test, it is not likely to be good production code. If we focus on the behavior we want, we will not be focusing on the structure we want.

Don't fool yourself. You cannot do both at once. It is hard enough to get code to behave the way you want it to. It is too hard to write it to behave *and* have the right structure. Thus, we follow Kent Beck's advice:

> *First make it work. Then make it right.*

Therefore, we add a new law to the three laws of TDD: the law of refactoring. First you write a small amount of failing test code. Then you write a small amount of passing production code. Then you clean up the mess you just made.

The traffic light gets a new color: red → green → refactor (Figure 2.1).

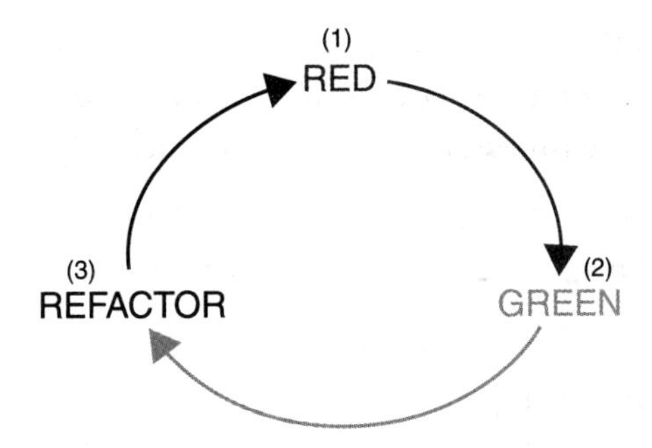

Figure 2.1 Red → green → refactor

You've likely heard of refactoring, and as I said earlier, we'll be spending a great deal of time on it in coming chapters. For now, let me dispel a few myths and misconceptions:

- Refactoring is a constant activity. Every time around the TDD cycle, you clean things up.
- Refactoring does not change behavior. You only refactor when the tests are passing, and the tests continue to pass while you refactor.
- Refactoring *never* appears on a schedule or a plan. You do not reserve time for refactoring. You do not ask permission to refactor. You simply refactor *all the time*.

Think of refactoring as the equivalent of washing your hands after using the restroom. It's just something you always do as a matter of common decency.

The Basics（基础知识）

It is very hard to create effective examples of TDD in text. The rhythm of TDD just doesn't come through very well. In the pages that follow, I try to convey that rhythm with appropriate timestamps and callouts. But to actually understand the true frequency of this rhythm, you just have to see it.

Therefore, each of the examples to follow has a corresponding online video that will help you see the rhythm first hand. Please watch each video in its entirety, and then make sure you go back to the text and read the explanation with the timestamps. If you don't have access to the videos, then pay special attention to the timestamps in the examples so you can infer the rhythm.

Simple Examples（简单示例）

As you review these examples, you are likely to discount them because they are all small and simple problems. You might conclude that TDD may be effective for such "toy examples" but cannot possibly work for complex systems. This would be a grave mistake.

The primary goal of any good software designer is to break down large and complex systems into a set of small, simple problems. The job of a programmer is to break those systems down into individual lines of code. Thus, the examples that follow are absolutely representative of TDD *regardless of the size of the project.*

This is something I can personally affirm. I have worked on large systems that were built with TDD, and I can tell you from experience that the rhythm and techniques of TDD are independent of scope. Size does not matter.

Or, rather, size does not matter to the procedure and the rhythm. However, size has a profound effect on the speed and coupling of the tests. But those are topics for the advanced chapters.

Stack（栈）

Watch related video: Stack
Access video by registering at informit.com/register

We start with a very simple problem: create a stack of integers. As we walk through this problem, note that the tests will answer any questions you have about the behavior of the stack. This is an example of the documentation value of tests. Note also that we appear to cheat by making the tests pass by plugging in absolute values. This is a common strategy in TDD and has a very important function. I'll describe that as we proceed.

We begin:

```
// T: 00:00 StackTest.java
package stack;

import org.junit.Test;

public class StackTest {
  @Test
  public void nothing() throws Exception {
  }
}
```

It's good practice to always start with a test that does nothing, and make sure that test passes. Doing so helps ensure that the execution environment is all working.

Next, we face the problem of what to test. There's no code yet, so what is there to test?

The answer to that question is simple. Assume we already know the code we want to write: `public class stack`. But we can't write it because we don't have a test that fails due to its absence. So, following the first law, we write the test that forces us to write the code that we already know we want to write.

> *Rule 1: Write the test that forces you to write the code you already know you want to write.*

This is the first of many rules to come. These "rules" are more like heuristics. They are bits of advice that I'll be throwing out, from time to time, as we progress through the examples.

Rule 1 isn't rocket science. If you can write a line of code, then you can write a test that tests that line of code, and you can write it first. Therefore,

```
// T:00:44 StackTest.java
public class StackTest {
  @Test
  public void canCreateStack() throws Exception {
    MyStack stack = new MyStack();
  }
}
```

I use **boldface** to show code that has been changed or added and highlight to depict code that does not compile. I chose `MyStack` for our example because the name `Stack` is known to the Java environment already.

Notice that in the code snippet, we changed the name of the test to communicate our intent. We can create a stack.

Now, because `MyStack` doesn't compile, we'd better follow the second law and create it, but by the third law, we'd better not write more than we need:

```
// T: 00:54 Stack.java
package stack;

public class MyStack {
}
```

Ten seconds have passed, and the test compiles and passes. When I initially wrote this example, most of that 10 seconds was taken up by rearranging my screen so that I can see both files at the same time. My screen now looks like Figure 2.2. The tests are on the left, and the production code is on the right. This is my typical arrangement. It's nice to have screen real estate.

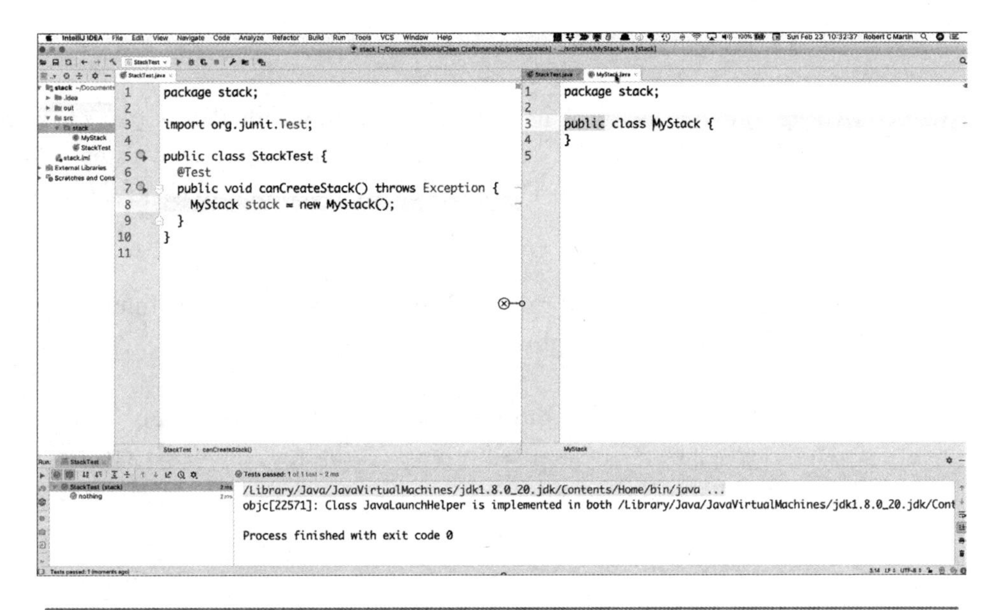

Figure 2.2 Rearranged screen

`MyStack` isn't a great name, but it avoided the name collision. Now that `MyStack` is declared in the `stack` package, let's change it back to `Stack`. That took 15 seconds. The tests still pass.

```
// T:01:09 StackTest.java
public class StackTest {
  @Test
  public void canCreateStack() throws Exception {
    Stack stack = new Stack();
  }
}
```

```
// T: 01:09 Stack.java
package stack;

public class Stack {
}
```

Here we see another rule: red → green → refactor. Never miss an opportunity to clean things up.

Rule 2: Make it fail. Make it pass. Clean it up.

Writing code that works is hard enough. Writing code that works and is clean is even harder. Fortunately, we can break the effort up into two steps. We can write bad code that works. Then, if we have tests, we can easily clean up the bad code while keeping it working.

Thus, every circuit around the TDD loop, we take the opportunity to clean up any little messes we might have made.

You may have noticed that our test does not actually assert any behavior. It compiles and passes but asserts nothing at all about the newly created stack. We can fix that in 15 seconds:

```
// T: 01:24 StackTest.java
public class StackTest {
  @Test
  public void canCreateStack() throws Exception {
    Stack stack = new Stack();
    assertTrue(stack.isEmpty());
  }
}
```

The second law kicks in here, so we'd better get this to compile:

```
// T: 01:49
import static junit.framework.TestCase.assertTrue;

public class StackTest {
  @Test
  public void canCreateStack() throws Exception {
    Stack stack = new Stack();
    assertTrue(stack.isEmpty());
  }
}
```

```
// T: 01:49 Stack.java
public class Stack {
  public boolean isEmpty() {
    return false;
  }
}
```

Twenty-five seconds later, it compiles but fails. The failure is intentional: `isEmpty` is specifically coded to return `false` because the first law says that the test must fail—but why does the first law demand this? Because now we can see that our test fails when it ought to fail. We have tested our test. Or rather, we have tested one half of it. We can test the other half by changing `isEmpty` to return `true`:

```
// T: 01:58 Stack.java
public class Stack {
  public boolean isEmpty() {
    return true;
  }
}
```

Nine seconds later, the test passes. It has taken *9 seconds* to ensure that the test both passes and fails.

When programmers first see that `false` and then that `true`, they often laugh because it looks so foolish. It looks like cheating. But it's not cheating, and it's not at all foolish. It has taken mere seconds to ensure that the test both passes and fails as it should. Why in the world would we *not* do this?

What's the next test? Well, we know we need to write the `push` function. So, by Rule 1, we write the test that forces us to write the `push` function:

```
// T 02:24 StackTest.java
@Test
public void canPush() throws Exception {
  Stack stack = new Stack();
  stack.push(0);
}
```

This doesn't compile. By the second law, then, we must write the production code that will make it compile:

```
// T: 02:31 Stack.java
public void push(int element) {

}
```

This compiles, of course, but now we have another test without an assertion. The obvious thing to assert is that, after one push, the stack is not empty:

```
// T: 02:54 StackTest.java
@Test
public void canPush() throws Exception {
  Stack stack = new Stack();
  stack.push(0);
  assertFalse(stack.isEmpty());
}
```

This fails, of course, because `isEmpty` returns `true`, so we need to do something a bit more intelligent—like create a Boolean flag to track emptiness:

```
// T: 03:46 Stack.java
public class Stack {
  private boolean empty = true;

  public boolean isEmpty() {
    return empty;
  }

  public void push(int element) {
    empty=false;
  }
}
```

This passes. It has been 2 minutes since the last test passed. Now, by Rule 2, we need to clean this up. The duplication of the stack creation bothers me, so let's extract the stack into a field of the class and initialize it:

```
// T: 04:24 StackTest.java
public class StackTest {
  private Stack stack = new Stack();

  @Test
  public void canCreateStack() throws Exception {
    assertTrue(stack.isEmpty());
  }

  @Test
  public void canPush() throws Exception {
    stack.push(0);
    assertFalse(stack.isEmpty());
  }
}
```

This requires 30 seconds, and the tests still pass.

The name `canPush` is a pretty bad name for this test.

```
// T: 04:50 StackTest.java
@Test
public void afterOnePush_isNotEmpty() throws Exception {
  stack.push(0);
  assertFalse(stack.isEmpty());
}
```

That's better. And, of course, it still passes.

Okay, back to the first law. If we push once and pop once, the stack should be empty again:

```
// T: 05:17 StackTest.java
@Test
```

```
public void afterOnePushAndOnePop_isEmpty() throws Exception {
  stack.push(0);
  stack.pop()
}
```

The second law kicks in because pop doesn't compile, so

```
// T: 05:31 Stack.java
public int pop() {
  return -1;
}
```

And then the third law allows us to finish the test:

```
// T: 05:51
@Test
public void afterOnePushAndOnePop_isEmpty() throws Exception {
  stack.push(0);
  stack.pop();
  assertTrue(stack.isEmpty());
}
```

This fails because nothing sets the `empty` flag back to true, so

```
// T: 06:06 Stack.java
public int pop() {
  empty=true;
  return -1;
}
```

And, of course, this passes. It has been 76 seconds since the last test passed.

Nothing to clean up, so back to the first law. The size of the stack should be 2 after two pushes.

```
// T: 06:48 StackTest.java
@Test
public void afterTwoPushes_sizeIsTwo() throws Exception {
```

```
    stack.push(0);
    stack.push(0);
    assertEquals(2, stack.getSize());
}
```

The second law kicks in because of the compile errors, but they are easy to fix. We add the necessary `import` to the test and the following function to the production code:

```
// T: 07:23 Stack.java
public int getSize() {
  return 0;
}
```

And now everything compiles, but the tests fail.

Of course, getting the test to pass is trivial:

```
// T: 07:32 Stack.java
public int getSize() {
  return 2;
}
```

This looks dumb, but we have now seen the test fail and pass properly, and it took only 11 seconds. So, again, why wouldn't we do this?

But this solution is clearly naive, so by Rule 1, we modify a previous test in a manner that will force us to write a better solution. And, of course, we screw it up (you can blame me):

```
// T: 08:06 StackTest.java
@Test
public void afterOnePushAndOnePop_isEmpty() throws Exception {
  stack.push(0);
  stack.pop();
  assertTrue(stack.isEmpty());
  assertEquals(1, stack.getSize());
}
```

Okay, that was really stupid. But programmers make dumb mistakes from time to time, and I am no exception. I didn't spot this mistake right away when I first wrote the example because the test failed just the way I expected it to.

So, now, secure in the assumption that our tests are good, let's make the changes that we believe will make those tests pass:

```
// T: 08:56
public class Stack {
  private boolean empty = true;
  private int size = 0;

  public boolean isEmpty() {
    return size == 0;
  }

  public void push(int element) {
    size++;
  }

  public int pop() {
    --size;
    return -1;
  }

  public int getSize() {
    return size;
  }
}
```

I was surprised to see this fail. But after composing myself, I quickly found my error and repaired the test. Let's do this:

```
// T: 09:28 StackTest.java
@Test
public void afterOnePushAndOnePop_isEmpty() throws Exception {
  stack.push(0);
```

```
  stack.pop();
  assertTrue(stack.isEmpty());
  assertEquals(0, stack.getSize());
}
```

And the tests all pass. It has been 3 minutes and 22 seconds since the tests last passed.

For the sake of completeness, let's add the size check to another test:

```
// T: 09:51 StackTest.java
@Test
public void afterOnePush_isNotEmpty() throws Exception {
  stack.push(0);
  assertFalse(stack.isEmpty());
  assertEquals(1, stack.getSize());
}
```

And, of course, that passes.

Back to the first law. What should happen if we pop an empty stack? We should expect an underflow exception:

```
// T: 10:27 StackTest.java
@Test(expected = Stack.Underflow.class)
public void poppingEmptyStack_throwsUnderflow() {
}
```

The second law forces us to add that exception:

```
// T: 10:36 Stack.java
public class Underflow extends RuntimeException {
}
```

And then we can complete the test:

```
// T: 10:50 StackTest.java
@Test(expected = Stack.Underflow.class)
```

```
public void poppingEmptyStack_throwsUnderflow() {
  stack.pop();
}
```

This fails, of course, but it is easy to make it pass:

```
// T: 11:18 Stack.java
public int pop() {
  if (size == 0)
    throw new Underflow();
  --size;
  return -1;
}
```

That passes. It has been 1 minute and 27 seconds since the tests last passed.

Back to the first law. The stack should remember what was pushed. Let's try the simplest case:

```
// T: 11:49 StackTest.java

@Test
public void afterPushingX_willPopX() throws Exception {
  stack.push(99);
  assertEquals(99, stack.pop());
}
```

This fails because `pop` is currently returning `-1`. We make it pass by returning 99:

```
// T: 11:57 Stack.java
public int pop() {
  if (size == 0)
    throw new Underflow();
  --size;
  return 99;
}
```

This is obviously insufficient, so by Rule 1, we add enough to the test to force us to be a bit smarter:

```
// T: 12:18 StackTest.java
@Test
public void afterPushingX_willPopX() throws Exception {
  stack.push(99);
  assertEquals(99, stack.pop());
  stack.push(88);
  assertEquals(88, stack.pop());
}
```

This fails because we're returning 99. We make it pass by adding a field to record the last push:

```
// T: 12:50 Stack.java
public class Stack {
  private int size = 0;
  private int element;

  public void push(int element) {
    size++;
    this.element = element;
  }

  public int pop() {
    if (size == 0)
      throw new Underflow();
    --size;
    return element;
  }
}
```

This passes. It has been 92 seconds since the tests last passed.

At this point, you are probably pretty frustrated with me. You might even be shouting at these pages, demanding that I stop messing around and just write the damned stack. But actually, I've been following Rule 3.

Rule 3: Don't go for the gold.

When you first start TDD, the temptation is overwhelming to tackle the hard or interesting things first. Someone writing a stack would be tempted to test first-in-last-out (FILO) behavior first. This is called "going for the gold." By now, you have noticed that I have purposely avoided testing anything stack-like. I've been focusing on all the ancillary stuff around the outside of the stack, things like emptiness and size.

Why haven't I been going for the gold? Why does Rule 3 exist? Because when you go for the gold too early, you tend to miss all the details around the outside. Also, as you will soon see, you tend to miss the simplifying opportunities that those ancillary details provide.

Anyway, the first law has just kicked in. We need to write a failing test. And the most obvious test to write is FILO behavior:

```
// T: 13:36 StackTest.java
@Test
public void afterPushingXandY_willPopYthenX() {
  stack.push(99);
  stack.push(88);
  assertEquals(88, stack.pop());
  assertEquals(99, stack.pop());
}
```

This fails. Getting it to pass will require that we remember more than one value, so we should probably use an array. Let's rename the field to `elements` and turn it into an array:

```
// T: 13:51 Stack.java
public class Stack {
  private int size = 0;
  private int[] elements = new int[2];

  public void push(int element) {
    size++;
```

```
    this.elements = element;
  }

  public int pop() {
    if (size == 0)
      throw new Underflow();
    --size;
    return elements;
  }
}
```

Ah, but there are compile errors. Let's tackle them one by one. The `elements` variable in `push` needs brackets:

```
  // T: 14:01 Stack.java
  public void push(int element) {
    size++;
    this.elements[] = element;
  }
```

We need something to put into those braces. Hmmm. There's that `size++` from the previous snippets:

```
  // T: 14:07 Stack.java
  public void push(int element) {
    this.elements[size++] = element;
  }
```

There's the `elements` variable in `pop` too. It needs braces:

```
  // T: 14:13
  public int pop() {
    if (size == 0)
      throw new Underflow();
    --size;
    return elements[];
  }
```

Oh look! There's that nice --size we could put in there:

```
// T: 14:24
public int pop() {
  if (size == 0)
    throw new Underflow();
  return elements[--size];
}
```

And now the tests pass again. It has been 94 seconds since the tests last passed.

And we're done. Oh, there's more we could do. The stack holds only two elements and does not deal with overflows, but there's nothing left that I want to demonstrate in this example. So, consider those improvements an exercise for you, my reader.

It has taken 14 minutes and 24 seconds to write an integer stack from scratch. The rhythm you saw here was real and is typical. This is how TDD feels, regardless of the scope of the project.

Exercise

Implement a first-in-first-out queue of integers using the technique shown previously. Use a fixed-sized array to hold the integers. You will likely need two pointers to keep track of where the elements are to be added and removed. When you are done, you may find that you have implemented a circular buffer.

Prime Factors（质因数）

Watch related video: Prime Factors
Access video by registering at informit.com/register

The next example has a story and a lesson. The story begins in 2002 or thereabouts. I had been using TDD for a couple of years by that time and was learning Ruby. My son, Justin, came home from school and asked me for help with a homework problem. The homework was to find the prime factors of a set of integers.

I told Justin to try to work the problem by himself and that I would write a little program for him that would check his work. He retired to his room, and I set my laptop up on the kitchen table and started to ponder how to code the algorithm to find prime factors.

I settled on the obvious approach of using the sieve of Eratosthenes to generate a list of prime numbers and then dividing those primes into the candidate number. I was about to code it when a thought occurred to me: *What if I just start writing tests and see what happens?*

I began writing tests and making them pass, following the TDD cycle. And this is what happened.

First watch the video if you can. It will show you a lot of the nuances that I can't show in text. In the text that follows, I avoid the tedium of all the timestamps, compile-time errors, and so on. I just show you the incremental progress of the tests and the code.

We begin with the most obvious and degenerate case. Indeed, that follows a rule:

> *Rule 4: Write the simplest, most specific, most degenerate*[2] *test that will fail.*

The most degenerate case is the prime factors of 1. The most degenerate failing solution is to simply return a `null`.

```
public class PrimeFactorsTest {
  @Test
  public void factors() throws Exception {
    assertThat(factorsOf(1), is(empty()));
  }
```

2. The word *degenerate* is used here to mean the most absurdly simple starting point.

```
  private List<Integer> factorsOf(int n) {
    return null;
  }
}
```

Note that I am including the function being tested within the test class. This is not typical but is convenient for this example. It allows me to avoid bouncing back and forth between two source files.

Now this test fails, but it is easy to make it pass. We simply return an empty list:

```
private List<Integer> factorsOf(int n) {
  return new ArrayList<>();
}
```

Of course, this passes. The next most degenerate test is 2.

```
assertThat(factorsOf(2), contains(2));
```

This fails; but again, it's easy to make it pass. That's one of the reasons we choose degenerate tests: they are almost always easy to make pass.

```
private List<Integer> factorsOf(int n) {
  ArrayList<Integer> factors = new ArrayList<>();
  if (n>1)
    factors.add(2);
  return factors;
}
```

If you watched the video, you saw that this was done in two steps. The first step was to extract the `new ArrayList<>()` as a variable named `factors`. The second step was to add the `if` statement.

I emphasize these two steps because the first follows Rule 5.

Rule 5: Generalize where possible.

The original constant, `new ArrayList<>()`, is very specific. It can be generalized by putting it into a variable that can be manipulated. It's a small generalization, but small generalizations often are all that are necessary.

And so, the tests pass again. The next most degenerate test elicits a fascinating result:

```
assertThat(factorsOf(3), contains(3));
```

This fails. Following Rule 5, we need to generalize. There is a very simple generalization that makes this test pass. It might surprise you. You'll have to look closely; otherwise you'll miss it.

```
private List<Integer> factorsOf(int n) {
  ArrayList<Integer> factors = new ArrayList<>();
  if (n>1)
    factors.add(n);
  return factors;
}
```

I sat at my kitchen table and marveled that a simple one-character change that replaced a constant with a variable, a simple generalization, made the new test pass and kept all previous tests passing.

I'd say we were on a roll, but the next test is going to be disappointing. The test itself is obvious:

```
assertThat(factorsOf(4), contains(2, 2));
```

But how do we solve that by generalizing? I can't think of a way. The only solution I can think of is to test whether `n` is divisible by 2, and that just isn't very general. Nevertheless,

```
private List<Integer> factorsOf(int n) {
  ArrayList<Integer> factors = new ArrayList<>();
  if (n>1) {
    if (n%2 == 0) {
      factors.add(2);
      n /= 2;
    }
    factors.add(n);
  }
  return factors;
}
```

Not only is this not very general; it also fails a previous test. It fails the test for the prime factors of 2. The reason should be clear. When we reduce `n` by a factor of 2, it becomes 1, which then gets put into the list.

We can fix that with some even less general code:

```
private List<Integer> factorsOf(int n) {
  ArrayList<Integer> factors = new ArrayList<>();
  if (n > 1) {
    if (n % 2 == 0) {
      factors.add(2);
      n /= 2;
    }
    if (n > 1)
      factors.add(n);
  }
  return factors;
}
```

At this point, you might fairly accuse me of just tossing in various `if` statements to make the tests pass. That's not far from the truth. You might also accuse me of violating Rule 5, because none of this recent code is particularly general. On the other hand, I don't see any options.

But there's a hint of a generalization to come. Notice that the two `if` statements have identical predicates. It's almost as if they were part of an unwound loop. Indeed, there's no reason that second `if` statement needs to be inside the first.

```
private List<Integer> factorsOf(int n) {
  ArrayList<Integer> factors = new ArrayList<>();
  if (n > 1) {
    if (n % 2 == 0) {
      factors.add(2);
      n /= 2;
    }
  }
  if (n > 1)
    factors.add(n);
  return factors;
}
```

This passes and looks *very* much like an unwound loop.

The next three tests pass without any changes:

```
assertThat(factorsOf(5), contains(5));
assertThat(factorsOf(6), contains(2,3));
assertThat(factorsOf(7), contains(7));
```

That's a pretty good indication that we are on the right track, and it makes me feel better about those ugly `if` statements.

The next most degenerate test is 8, and it must fail because our solution code simply cannot put three things into the list:

```
assertThat(factorsOf(8), contains(2, 2, 2));
```

The way to make this pass is another surprise—and a powerful application of Rule 5. We change an `if` to a `while`:

```
private List<Integer> factorsOf(int n) {
  ArrayList<Integer> factors = new ArrayList<>();
```

```
    if (n > 1) {
      while (n % 2 == 0) {
        factors.add(2);
        n /= 2;
      }
    }
    if (n > 1)
      factors.add(n);
    return factors;
  }
```

I sat at my kitchen table and once again I marveled. It seemed to me that something profound had happened here. At that time, I didn't know what it was. I do now. It was Rule 5. It turns out the `while` is a general form of `if`, and `if` is the degenerate form of `while`.

The next test, 9, must also fail because nothing in our solution factors out 3:

```
assertThat(factorsOf(9), contains(3, 3));
```

To solve it, we need to factor out 3s. We could do that as follows:

```
private List<Integer> factorsOf(int n) {
  ArrayList<Integer> factors = new ArrayList<>();
  if (n > 1) {
    while (n % 2 == 0) {
      factors.add(2);
      n /= 2;
    }
    while (n % 3 == 0) {
      factors.add(3);
      n /= 3;
    }
  }
  if (n > 1)
    factors.add(n);
  return factors;
}
```

But this is horrific. Not only is it a gross violation of Rule 5, but it's also a huge duplication of code. I'm not sure which rule violation is worse!

And this is where the *generalization mantra* kicks in:

> *As the tests get more specific, the code gets more generic.*

Every new test we write makes the test suite more specific. Every time we invoke Rule 5, the solution code gets more generic. We'll come back to this mantra later. It turns out to be critically important for test design and for the prevention of *fragile tests*.

We can eliminate the violation of duplication, and Rule 5, by putting the original factoring code into a loop:

```
private List<Integer> factorsOf(int n) {
  ArrayList<Integer> factors = new ArrayList<>();
  int divisor = 2;
  while (n > 1) {
    while (n % divisor == 0) {
      factors.add(divisor);
      n /= divisor;
    }
    divisor++;
  }
  if (n > 1)
    factors.add(n);
  return factors;
}
```

Once again, if you watch the video, you will see that this was done in several steps. The first step was to extract the three `2`s into the `divisor` variable. The next step was to introduce the `divisor++` statement. Then the initialization of the `divisor` variable was moved above the `if` statement. Finally, the `if` was changed into a `while`.

There it is again: that transition from `if->while`. Did you notice that the predicate of the original `if` statement turned out to be the predicate for the outer `while` loop? I found this to be startling. There's something genetic about it. It's as though the creature that I've been trying to create started from a simple seed and gradually evolved through a sequence of tiny mutations.

Notice that the `if` statement at the bottom has become superfluous. The only way the loop can terminate is if `n` is 1. That `if` statement really was the terminating condition of an unwound loop!

```
private List<Integer> factorsOf(int n) {
  ArrayList<Integer> factors = new ArrayList<>();
  int divisor = 2;
  while (n > 1) {
    while (n % divisor == 0) {
      factors.add(divisor);
      n /= divisor;
    }
    divisor++;
  }

  return factors;
}
```

Just a little bit of refactoring, and we get this:

```
private List<Integer> factorsOf(int n) {
  ArrayList<Integer> factors = new ArrayList<>();

  for (int divisor = 2; n > 1; divisor++)
    for (; n % divisor == 0; n /= divisor)
      factors.add(divisor);

  return factors;
}
```

And we're done. If you watch the video, you'll see that I added one more test, which proves that this algorithm is sufficient.

Sitting at my kitchen table, I saw those three salient lines and I had two questions. Where did this algorithm come from, and how does it work?

Clearly, it came from my brain. It was my fingers on the keyboard after all. But this was certainly not the algorithm I had planned on creating at the start. Where was the sieve of Eratosthenes? Where was the list of prime numbers? None of it was there!

Worse, why does this algorithm work? I was astounded that I could create a working algorithm and yet not understand how it functioned. I had to study it for a while to figure it out. My dilemma was the `divisor++` incrementer of the outer loop, which guarantees that every integer will be checked as a factor, including composite factors! Given the integer 12, that incrementer will check whether 4 is a factor. Why doesn't it put 4 in the list?

The answer is in the order of the execution, of course. By the time the incrementer gets to 4, all the 2s have been removed from `n`. And if you think about that for a bit, you'll realize that it *is* the sieve of Eratosthenes—but in a very different form than usual.

The bottom line here is that I derived this algorithm one test case at a time. I did not think it through up front. I didn't even know what this algorithm was going to look like when I started. The algorithm seemed to almost put itself together before my eyes. Again, it was like an embryo evolving one small step at a time into an ever-more-complex organism.

Even now, if you look at those three lines, you can see the humble beginnings. You can see the remnants of that initial `if` statement and fragments of all the other changes. The breadcrumbs are all there.

And we are left with a disturbing possibility. Perhaps TDD is a general technique for incrementally deriving algorithms. Perhaps, given a properly ordered suite of tests, we can use TDD to derive any computer program in a step-by-step, determinative, manner.

In 1936, Alan Turing and Alonzo Church separately proved that there was no general procedure for determining if there was a program for any given problem.[3] In so doing, they separately, and respectively, invented procedural and functional programming. Now TDD looks like it might be a general procedure for deriving the algorithms that solve the problems that *can* be solved.

The Bowling Game（保龄球局）

In 1999, Bob Koss and I were together at a C++ conference. We had some time to kill, so we decided to practice this new idea of TDD. We chose the simple problem of computing the score of a game of bowling.

A game of bowling consists of ten frames. In each frame, the player is given two attempts to roll a ball toward ten wooden pins in order to knock them down. The number of pins knocked down by a ball is the score for that ball. If all ten pins are knocked down by the first ball, it is called a *strike*. If knocking all ten pins down requires both balls, then it is called a *spare*. The dreaded gutter ball (Figure 2.3) yields no points at all.

Figure 2.3 The infamous gutter ball

3. This was Hilbert's "decidability problem." He asked whether there was a generalized way to prove that any given Diophantine equation was solvable. A Diophantine equation is a mathematical function with integer inputs and outputs. A computer program is also a mathematical function with integer inputs and outputs. Therefore, Hilbert's question can be described as pertaining to computer programs.

The scoring rules can be stated succinctly as follows:

- If the frame is a strike, the score is 10 plus the next two balls.
- If the frame is a spare, the score is 10 plus the next ball.
- Otherwise, the score is the two balls in the frame.

The score sheet in Figure 2.4 is a typical (if rather erratic) game.

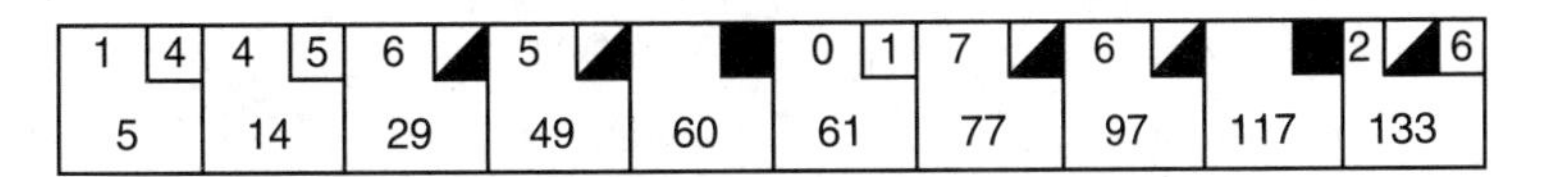

Figure 2.4 Score sheet of a typical game

On the player's first attempt, he knocked down one pin. On his second attempt, he knocked down four more, for a total of 5 points.

In the second frame, he rolled a 4 followed by a 5, giving him a 9 for the frame and a total of 14.

In the third frame, he rolled a 6 followed by a 4 (a spare). The total for that frame cannot be computed until the player begins the next frame.

In the fourth frame, the player rolls a 5. Now the score can be computed for the previous frame, which is 15, for a total of 29 in the third frame.

The spare in the fourth frame must wait until the fifth frame, for which the player rolls a strike. The fourth frame is therefore 20 points for a total of 49.

The strike in the fifth frame cannot be scored until the player rolls the next two balls for the sixth frame. Unfortunately, he rolls a 0 and a 1, giving him only 11 points for the fifth frame and a total of 60.

And on it goes until the tenth and final frame. Here the player rolls a spare and is thus allowed to roll one extra ball to finish out that spare.

Now, you are a programmer, a good object-oriented programmer. What are the classes and relationships you would use to represent the problem of computing the score of a game of bowling? Can you draw it in UML?[4]

Perhaps you might come up with something like what is shown in Figure 2.5.

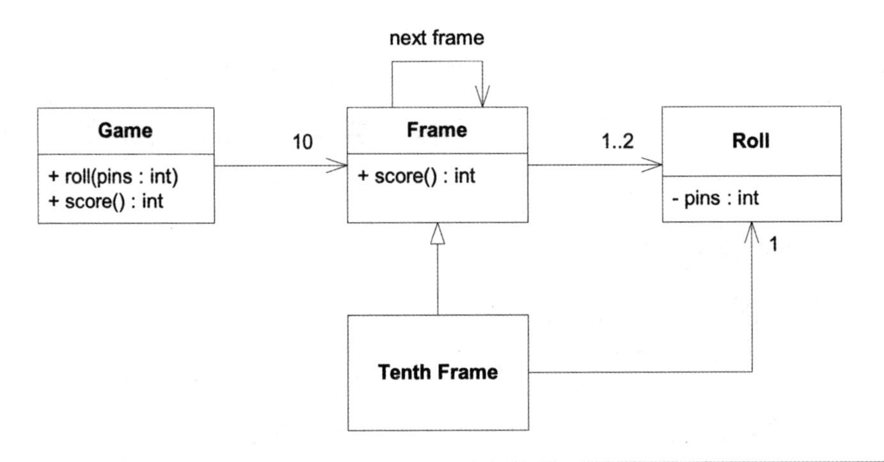

Figure 2.5 A UML diagram of the scoring of bowling

The `Game` has ten `Frames`. Each Frame has one or two `Rolls`, except the `TenthFrame` subclass, which inherits the `1..2` and adds one more roll to make it `2..3`. Each `Frame` object points to the next `Frame` so that the `score` function can look ahead in case it has to score a spare or a strike.

The `Game` has two functions. The `roll` function is called every time the player rolls a ball, and it is passed the number of pins the player knocked down. The `score` function is called once all the balls are rolled, and it returns the score for the entire game.

That's a nice, simple object-oriented model. It ought to be pretty easy to code. Indeed, given a team of four people, we could divide the work into the four classes and then meet a day or so later to integrate them and get them working.

4. The Unified Modeling Language. If you don't know UML, don't worry—it's just arrows and rectangles. You'll work it out from the description in the text.

Or, we could use TDD. If you can watch the video, you should do that now. In any case, please read through the text that follows.

Watch related video: Bowling Game
Access video by registering at informit.com/register

We begin, as always, with a test that does nothing, just to prove we can compile and execute. Once this test runs, we delete it:

```
public class BowlingTest {
  @Test
  public void nothing() throws Exception {
  }
}
```

Next, we assert that we can create an instance of the `Game` class:

```
@Test
public void canCreateGame() throws Exception {
  Game g = new Game();


}
```

Then we make that compile and pass by directing our IDE to create the missing class:

```
public class Game {
}
```

Next, we see if we can roll one ball:

```
@Test
public void canRoll() throws Exception {
  Game g = new Game();
  g.roll(0);
}
```

And then we make that compile and pass by directing the IDE to create the `roll` function, and we give the argument a reasonable name:

```
public class Game {
  public void roll(int pins) {
  }

}
```

You're probably already bored. This should be nothing new to you by now. But bear with me—it's about to get interesting. There's a bit of duplication in the tests already. We should get rid of it, so we factor out the creating of the game into the `setup` function:

```
public class BowlingTest {
  private Game g;

  @Before
  public void setUp() throws Exception {
    g = new Game();
  }
}
```

This makes the first test completely empty, so we delete it. The second test is also pretty useless because it doesn't assert anything. We delete it as well. Those tests served their purpose. They were stairstep tests.

> ***Stairstep tests:*** *Some tests are written just to force us to create classes or functions or other structures that we're going to need. Sometimes these tests are so degenerate that they assert nothing. Other times they assert something very naive. Often these tests are superseded by more comprehensive tests later and can be safely deleted. We call these kinds of tests stairstep tests because they are like stairs that allow us to incrementally increase the complexity to the appropriate level.*

Next, we want to assert that we can score a game. But to do that, we need to roll a complete game. Remember that the `score` function can be called only after all the balls in a game are rolled.

We fall back on Rule 4 and roll the simplest, most degenerate game we can think of:

```
@Test
public void gutterGame() throws Exception {
  for (int i=0; i<20; i++)
    g.roll(0);
  assertEquals(0, g.score());
}
```

Making this pass is trivial. We just need to return zero from `score`. But first we return `-1` (not shown) just to see it fail. Then we return zero to see it pass:

```
public class Game {
  public void roll(int pins) {
  }

  public int score() {
    return 0;
  }
}
```

Okay, I said this was about to get interesting, and it is. Just one more little setup. The next test is another example of Rule 4. The next most degenerate test I can think of is all ones. We can write this test with a simple copy paste from the last test:

```
@Test
public void allOnes() throws Exception {
  for (int i=0; i<20; i++)
    g.roll(1);
  assertEquals(20, g.score());
}
```

This creates some duplicate code. The last two tests are virtually identical. When we refactor, we'll have to fix that. But first, we need to make this test pass, and that's really easy. All we need to do is add up all the rolls:

```
public class Game {
  private int score;

  public void roll(int pins) {
    score += pins;
  }

  public int score() {
    return score;
  }
}
```

Of course, this is not the right algorithm for scoring bowling. Indeed, it is hard to see how this algorithm could ever evolve into becoming the rules for scoring bowling. So, I'm suspicious—I expect squalls in the coming tests. But for now, we must refactor.

The duplication in the tests can be eliminated by extracting a function called `rollMany`. The IDE's *Extract Method* refactoring helps immensely with this and even detects and replaces both instances of the duplication:

```
public class BowlingTest {
  private Game g;

  @Before
  public void setUp() throws Exception {
    g = new Game();
  }

  private void rollMany(int n, int pins) {
    for (int i = 0; i < n; i++) {
      g.roll(pins);
    }
  }
```

```
  @Test
  public void gutterGame() throws Exception {
    rollMany(20, 0);
    assertEquals(0, g.score());
  }

  @Test
  public void allOnes() throws Exception {
    rollMany(20, 1);
    assertEquals(20, g.score());
  }
}
```

Okay, next test. It's hard to think of something degenerate at this point, so we might as well try for a spare. We'll keep it simple though: one spare, with one extra bonus ball, and all the rest gutter balls.

```
@Test
public void oneSpare() throws Exception {
  rollMany(2, 5); // spare
  g.roll(7);
  rollMany(17, 0);
  assertEquals(24, g.score());
}
```

Let's check my logic: This game has two balls in each frame. The first two balls are the spare. The next ball is the ball after the spare, and the 17 gutter balls complete the game.

The score in the first frame is 17, which is 10 plus the 7 rolled in the next frame. The score for the whole game, therefore, is 24 because the 7 is counted twice. Convince yourself that this is correct.

This test fails, of course. So how do we get it to pass? Let's look at the code:

```
public class Game {
  private int score;
```

```
  public void roll(int pins) {
    score += pins;
  }

  public int score() {
    return score;
  }
}
```

The score is being calculated in the `roll` function, so we need to modify that function to account for the spare. But that will force us to do something really ugly, like this:

```
public void roll(int pins) {
   if (pins + lastPins == 10) { // horrors!
     //God knows what...
   }
   score += pins;
  }
```

That `lastPins` variable must be a field of the `Game` class that remembers what the last roll was. And if the last roll and this roll add up to 10, then that's a spare. Right? Ew!

You should feel your sphincter muscles tightening. You should feel your gorge rise and a tension headache beginning to build. The angst of the software craftsman should be raising your blood pressure.

This is just wrong!

We've all had that feeling before, haven't we? The question is, what do you do about it?

Whenever you get that feeling that something is wrong, trust it! So, what is it that's wrong?

There's a design flaw. You might rightly wonder how there could be a design flaw in two executable lines of code. But the design flaw is there; it's blatant, and it's deeply harmful. As soon as I tell you what it is you'll recognize it and agree with me. Can you find it?

Jeopardy song interlude.

I told you what the design flaw was at the very start. Which of the two functions in this class claims, *by its name*, to calculate the score? The `score` function, of course. Which function actually does calculate the score? The `roll` function. That's misplaced responsibility.

> ***Misplaced responsibility:*** *A design flaw in which the function that claims to perform a computation does not actually perform the computation. The computation is performed elsewhere.*

How many times have you gone to the function that claims to do some task, only to find that function does not do that task? And now you have no idea where, in the system, that task is actually done. Why does this happen?

Clever programmers. Or rather, programmers who *think* they are clever.

It was very clever of us to sum up the pins in the `roll` function, wasn't it? We knew that function was going to be called once per roll, and we knew that all we had to do was sum up the rolls, so we just put that addition right there. Clever, clever, clever. And that cleverness leads us to Rule 6.

> *Rule 6: When the code feels wrong, fix the design before proceeding.*

So how do we fix this design flaw? The calculation of the score is in the wrong place, so we're going to have to move it. By moving it, perhaps we'll be able to figure out how to pass the spare test.

Moving the calculation means that the `roll` function is going to have to remember all the rolls in something like an array. Then the `score` function can sum up the array.

```
public class Game {
  private int rolls[] = new int[21];
  private int currentRoll = 0;

  public void roll(int pins) {
    rolls[currentRoll++] = pins;
  }

  public int score() {
    int score = 0;
    for (int i = 0; i < rolls.length; i++) {
      score += rolls[i];
    }
    return score;
  }
}
```

This fails the spare test, but it passes the other two tests. What's more, the spare test fails for the same reason as before. So, although we've completely changed the structure of the code, the behavior remains unchanged. By the way, that is the *definition* of refactoring.

> ***Refactoring:*** *A change to the structure of the code that has no effect upon the behavior.*[5]

Can we pass the spare case now? Well, maybe, but it's still icky:

```
public int score() {
  int score = 0;
  for (int i = 0; i < rolls.length; i++) {
    if (rolls[i] + rolls[i+1] == 10) { // icky
      // What now?
    }
    score += rolls[i];
  }
  return score;
}
```

5. Martin Fowler, *Refactoring: Improving the Design of Existing Code*, 2nd ed. (Addison-Wesley, 2019).

Is that right? No, it's wrong, isn't it? It only works if `i` is even. To make that `if` statement actually detect a spare, it would have to look like this:

```
if (rolls[i] + rolls[i+1] == 10 && i%2 == 0) { // icky
```

So, we're back to Rule 6—there's another design problem. What could it be?

Look back to the UML diagram earlier in this chapter. That diagram shows that the `Game` class should have ten `Frame` instances. Is there any wisdom in that? Look at our loop. At the moment, it's going to loop 21 times! Does that make any sense?

Let me put it this way. If you were about to review the code for scoring bowling—code that you'd never seen before—what number would you expect to see in that code? Would it be 21? Or would it be 10?

I hope you said 10, because there are 10 frames in a game of bowling. Where is the number 10 in our scoring algorithm? Nowhere!

How can we get the number 10 into our algorithm? We need to loop through the array *one frame at a time*. How do we do that?

Well, we could loop through the array two balls at a time, couldn't we? I mean like this:

```
public int score() {
  int score = 0;
  int i = 0;
  for (int frame = 0; frame<10; frame++) {
    score += rolls[i] + rolls[i+1];
    i += 2;
  }
  return score;
}
```

Once again, this passes the first two tests and fails the spare test for the same reason as before. So, no behavior was changed. This was a true refactoring.

You might be ready to tear this book up now because you know that looping through the array two balls at a time is just plain wrong. Strikes have only one ball in their frame, and the tenth frame could have three.

True enough. However, so far none of our tests has used a strike, or the tenth frame. So, for the moment, two balls per frame works fine.

Can we pass the spare case now? Yes. It's trivial:

```
public int score() {
  int score = 0;
  int i = 0;
  for (int frame = 0; frame < 10; frame++) {
    if (rolls[i] + rolls[i + 1] == 10) { // spare
     score += 10 + rolls[i + 2];
      i += 2;
    } else {
      score += rolls[i] + rolls[i + 1];
      i += 2;
    }
  }
  return score;
}
```

This passes the spare test. Nice. But the code is pretty ugly. We can rename `i` to `frameIndex` and get rid of that ugly comment by extracting a nice little method:

```
public int score() {
  int score = 0;
  int frameIndex = 0;
  for (int frame = 0; frame < 10; frame++) {
    if (isSpare(frameIndex)) {
      score += 10 + rolls[frameIndex + 2];
```

```
        frameIndex += 2;
      } else {
        score += rolls[frameIndex] + rolls[frameIndex + 1];
        frameIndex += 2;
      }
    }
    return score;
  }

  private boolean isSpare(int frameIndex) {
    return rolls[frameIndex] + rolls[frameIndex + 1] == 10;
  }
```

That's better. We can also clean up the ugly comment in the spare test by doing the same:

```
private void rollSpare() {
  rollMany(2, 5);
}

@Test
public void oneSpare() throws Exception {
  rollSpare();
  g.roll(7);
  rollMany(17, 0);
  assertEquals(24, g.score());
}
```

Replacing comments with pleasant little functions like this is almost always a good idea. The folks who read your code later will thank you.

So, what's the next test? I suppose we should try a strike:

```
@Test
public void oneStrike() throws Exception {
  g.roll(10); // strike
  g.roll(2);
  g.roll(3);
  rollMany(16, 0);
```

```
    assertEquals(20, g.score());
  }
```

Convince yourself that this is correct. There's the strike, the 2 bonus balls, and 16 gutter balls to fill up the remaining eight frames. The score is 15 in the first frame and 5 in the second. All the rest are 0 for a total of 20.

This test fails, of course. What do we have to do to make it pass?

```
public int score() {
  int score = 0;
  int frameIndex = 0;
  for (int frame = 0; frame < 10; frame++) {
    if (rolls[frameIndex] == 10) { // strike
      score += 10 + rolls[frameIndex+1] +
                    rolls[frameIndex+2];
      frameIndex++;
    }
    else if (isSpare(frameIndex)) {
      score += 10 + rolls[frameIndex + 2];
      frameIndex += 2;
    } else {
      score += rolls[frameIndex] + rolls[frameIndex + 1];
      frameIndex += 2;
    }
  }
  return score;
}
```

This passes. Note that we increment `frameIndex` only by one. That's because a strike has only one ball in a frame—and you were so worried about that, weren't you?

This is a very good example of what happens when you get the design right. The rest of the code just starts trivially falling into place. Pay special attention to Rule 6, boys and girls, and get the design right early. It will save you immense amounts of time.

We can clean this up quite a bit. That ugly comment can be fixed by extracting an `isStrike` method. We can extract some of that ugly math into some pleasantly named functions too. When we're done, it looks like this:

```
public int score() {
  int score = 0;
  int frameIndex = 0;
  for (int frame = 0; frame < 10; frame++) {
    if (isStrike(frameIndex)) {
      score += 10 + strikeBonus(frameIndex);
      frameIndex++;
    } else if (isSpare(frameIndex)) {
      score += 10 + spareBonus(frameIndex);
      frameIndex += 2;
    } else {
      score += twoBallsInFrame(frameIndex);
      frameIndex += 2;
    }
  }
  return score;
}
```

We can also clean up the ugly comment in the test by extracting a `rollStrike` method:

```
@Test
public void oneStrike() throws Exception {
  rollStrike();
  g.roll(2);
  g.roll(3);
  rollMany(16, 0);
  assertEquals(20, g.score());
}
```

What's the next test? We haven't tested the tenth frame yet. But I'm starting to feel pretty good about this code. I think it's time to break Rule 3 and *go for the gold*. Let's test a perfect game!

```
@Test
public void perfectGame() throws Exception {
  rollMany(12, 10);
  assertEquals(300, g.score());
}
```

We roll a strike in the first nine frames, and then a strike and two 10s in the tenth frame. The score, of course, is 300—everybody knows that.

What's going to happen when I run this test? It should fail, right? But, no! It passes! It passes because we are done! The `score` function *is* the solution. You can prove that to yourself by reading it. Here, you follow along while I read it to you:

```
For each of the ten frames
   If that frame is a strike,
     Then the score is 10 plus the strike bonus
      (the next two balls).
   If that frame is a spare,
     Then the score is 10 plus the spare bonus
      (the next ball).
   Otherwise,
     The score is the two balls in the frame.
```

The code reads like the rules for scoring bowling. Go to the beginning of this chapter and read those rules again. Compare them to the code. And then ask yourself if you've ever seen requirements and code that are so closely aligned.

Some of you may be confused about why this works. You look at the tenth frame on that scorecard and you see that it doesn't look like any of the other frames; and yet there is no code in our solution that makes the tenth frame a special case. How can that be?

The answer is that the tenth frame is not special at all. It is drawn differently on the score card, but it is not scored differently. There is no special case for the tenth frame.

And we were going to make a subclass out of it!

Look back at that UML diagram. We could have divvied out the tasks to three or four programmers and integrated a day or two later. And the tragedy is that we'd have gotten it working. We would have celebrated the 400[6] lines of working code, never knowing that the algorithm was a `for` loop and two `if` statements that fits into 14 lines of code.

Did you see the solution early? Did you see the `for` loop and two `if` statements? Or did you expect one of the tests to eventually force me to write the `Frame` class? Were you holding out for the tenth frame? Is that where you thought all the complexity would be found?

Did you know we were done before we ran the tenth frame test? Or did you think there was a lot more to do? Isn't it fascinating that we can write a test fully expecting more work ahead, only to find, to our surprise, that we are done?

Some folks have complained that if we'd followed the UML diagram shown earlier, we'd have wound up with code that was easier to change and maintain. That's pure bollocks! What would you rather maintain, 400 lines of code in four classes or 14 lines with a `for` loop and two `if` statements?

Conclusion（小结）

In this chapter, we studied the motivations and basics of TDD. If you've gotten this far, your head may be spinning. We covered a lot of ground. But not nearly enough. The next chapter goes significantly deeper into the topic of TDD, so you may wish to take a little rest before turning the page.

6. I know it's 400 lines of code because I've written it.

3 Advanced TDD

（高级测试驱动开发）

Hold on to your hats. The ride is about to get fast and bumpy. To quote Dr. Morbius as he led a tour of the Krell machine: "Prepare your minds for a new scale of scientific values."

SORT 1（排序示例一）

The last two examples in Chapter 2, "Test-Driven Development," begged an interesting question. Where does the algorithm that we derive using test-driven development (TDD) come from? Clearly it comes from our brain, but not in the way we are used to. Somehow, the sequence of failing tests coaxes the algorithm out of our brain without the need to think it all through beforehand.

This raises the possibility that TDD may be a step-by-step, incremental procedure for deriving any algorithm for any problem. Think of it like solving a mathematical or geometric proof. You start from basic postulates—the degenerate failing tests. Then, one step at a time, you build up the solution to the problem.

At each step, the tests get more and more constraining and specific, but the production code gets more and more generic. This process continues until the production code is so general that you can't think of any more tests that will fail. That solves the entire problem.

Let's try this out. Let's use the approach to derive an algorithm for sorting an array of integers.

If you can watch the video, this would be a good time, but in any case, please continue on to read the text that follows.

Watch related video: SORT 1
Access video by registering at informit.com/register

We begin, as usual, with a test that does nothing:

```
public class SortTest {
  @Test
  public void nothing() throws Exception {

  }
}
```

The first failing test will be the degenerate case of an empty array:

```
public class SortTest {

  @Test
  public void sorted() throws Exception {
    assertEquals(asList(), sort(asList()));
  }
  private List<Integer> sort(List<Integer> list) {
    return null;
  }
}
```

This obviously fails, but it is easy to make it pass:

```
private List<Integer> sort(List<Integer> list) {
  return new ArrayList<>();
}
```

Stepping up one level in degeneracy, we try a list with one integer in it:

```
assertEquals(asList(1), sort(asList(1)));
```

This clearly fails, but we can make it pass by making the production code a bit more general:

```
private List<Integer> sort(List<Integer> list) {
  return list;
}
```

Cute, right? And we saw that trick before, in the prime factors example in Chapter 2. It seems relatively common that the first two tests in a given problem are solved by returning the most degenerate answer, followed by the input argument.

The next test is trivial because it already passes: two elements in order. It could be argued that we shouldn't even have written it because it's not a failing test. But it's nice to see these tests pass.

```
assertEquals(asList(1, 2), sort(asList(1, 2)));
```

If we reverse the order of the input array, the test will fail: two elements out of order.

```
assertEquals(asList(1, 2), sort(asList(2, 1)));
```

To make this pass, we're going to have to do something marginally intelligent. If the input array has more than one element in it and the first two elements of the array are out of order, then we should swap them:

```
private List<Integer> sort(List<Integer> list) {
  if (list.size() > 1) {
    if (list.get(0) > list.get(1)) {
      int first = list.get(0);
      int second = list.get(1);
      list.set(0, second);
      list.set(1, first);
    }
  }
  return list;
}
```

Maybe you can see where this is headed. If so, don't spoil the surprise for everyone else. Also, remember this moment—we'll be returning to it in the next section.

The next two tests pass already. In the first test, the input array is already in order. In the second test, the first two elements are out of order, and our current solution swaps them.

```
assertEquals(asList(1, 2, 3), sort(asList(1, 2, 3)));
assertEquals(asList(1, 2, 3), sort(asList(2, 1, 3)));
```

The next failing test is three elements with the second two out of order:

```
assertEquals(asList(1, 2, 3), sort(asList(2, 3, 1)));
```

We get this one to pass by putting our compare and swap algorithm into a loop that walks down the length of the list:

```
private List<Integer> sort(List<Integer> list) {
  if (list.size() > 1) {
    for (int firstIndex=0; firstIndex < list.size()-1; firstIndex++) {
      int secondIndex = firstIndex + 1;
      if (list.get(firstIndex) > list.get(secondIndex)) {
        int first = list.get(firstIndex);
        int second = list.get(secondIndex);
        list.set(firstIndex, second);
        list.set(secondIndex, first);
      }
    }
  }
  return list;
}
```

Can you tell where this is going yet? Most of you likely do. Anyway, the next failing test case is three elements in reverse order:

```
assertEquals(asList(1, 2, 3), sort(asList(3, 2, 1)));
```

The failure results are telling. The `sort` function returns `[2, 1, 3]`. Note that the `3` got moved all the way to the end of the list. That's good! But the first two elements are still out of order. It's not hard to see why. The `3` got swapped with the `2`, and then the `3` got swapped with the `1`. But that left the `2` and `1` still out of order. They need to be swapped again.

So, the way to get this test to pass is to put the compare and swap loop into another loop that incrementally reduces the length of the comparing and swapping. Maybe that's easier to read in code:

```
private List<Integer> sort(List<Integer> list) {
  if (list.size() > 1) {
    for (int limit = list.size() - 1; limit > 0; limit--) {
      for (int firstIndex = 0; firstIndex < limit; firstIndex++) {
        int secondIndex = firstIndex + 1;
        if (list.get(firstIndex) > list.get(secondIndex)) {
          int first = list.get(firstIndex);
          int second = list.get(secondIndex);
          list.set(firstIndex, second);
          list.set(secondIndex, first);
        }
      }
    }
  }
  return list;
}
```

To finish this off, let's do a larger-scale test:

```
assertEquals(
             asList(1, 1, 2, 3, 3, 3, 4, 5, 5, 5, 6, 7, 8, 9, 9, 9),
             sort(asList(3, 1, 4, 1, 5, 9, 2, 6, 5, 3, 5, 8, 9, 7, 9,
                    3)));
```

This passes, so our sort algorithm appears to be complete.

Where did this algorithm come from? We did not design it up front. It just came from the set of small decisions we made in order to get each failing test to pass. It was an incremental derivation. Voila!

And what algorithm is this? *Bubble sort*, of course—one of the worst possible sorting algorithms.

So maybe TDD is a really good way to incrementally derive really bad algorithms.

Sort 2（排序示例二）

Let's try this again. This time we'll choose a slightly different pathway. And again, view the video if possible, but then continue reading from this point.

Watch related video: SORT 2
Access video by registering at informit.com/register

We begin as we did before with the most degenerate tests possible and the code that makes them pass:

```
public class SortTest {
  @Test
  public void testSort() throws Exception {
    assertEquals(asList(), sort(asList()));
    assertEquals(asList(1), sort(asList(1)));
    assertEquals(asList(1, 2), sort(asList(1, 2)));
  }

  private List<Integer> sort(List<Integer> list) {
    return list;
  }
}
```

As before, we then pose two items out of order:

```
assertEquals(asList(1, 2), sort(asList(2, 1)));
```

But now, instead of comparing and swapping them within the input `list`, we compare and create an entirely new list with the elements in the right order:

```
private List<Integer> sort(List<Integer> list) {
  if (list.size() <= 1)
    return list;
  else {
    int first  = list.get(0);
    int second = list.get(1);
    if (first > second)
      return asList(second, first);
    else
      return asList(first, second);
  }
}
```

This is a good moment to pause and reflect. When we faced this test in the previous section, we blithely wrote the compare and swap solution as though it were the only possible way to pass the test. But we were wrong. In this example, we see another way.

This tells us that from time to time we may encounter failing tests that have more than one possible solution. Think of these as forks in the road. Which fork should we take?

Let's watch how this fork proceeds.

The obvious next test is, as it was before, three elements in order:

```
assertEquals(asList(1, 2, 3), sort(asList(1, 2, 3)));
```

But unlike in the previous example, this test fails. It fails because none of the pathways through the code can return a list with more than two elements. However, making it pass is trivial:

```
private List<Integer> sort(List<Integer> list) {
  if (list.size() <= 1)
```

```
    return list;
  else if (list.size() == 2){
    int first = list.get(0);
    int second = list.get(1);
    if (first > second)
      return asList(second, first);
    else
      return asList(first, second);
  }
  else {
    return list;
  }
}
```

Of course, this is silly, but the next test—three elements with the first two out of order—eliminates the silliness. This obviously fails:

```
assertEquals(asList(1, 2, 3), sort(asList(2, 1, 3)));
```

How the devil are we going to get this to pass? There are only two possibilities for a list with two elements, and our solution exhausts both of those possibilities. But with three elements, there are *six* possibilities. Are we really going to decode and construct all six possible combinations?

No, that would be absurd. We need a simpler approach. What if we use the law of trichotomy?

The law of trichotomy says that given two numbers A and B, there are only three possible relationships between them: $A < B$, $A = B$, or $A > B$. Okay, so let's arbitrarily pick one of the elements of the list and then decide which of those relationships it has with the others.

The relevant code looks like this:

```
else {
  int first = list.get(0);
  int middle = list.get(1);
```

```
    int last = list.get(2);
    List<Integer> lessers = new ArrayList<>();
    List<Integer> greaters = new ArrayList<>();

    if (first < middle)
      lessers.add(first);
    if (last < middle)
      lessers.add(last);
    if (first > middle)
      greaters.add(first);
    if (last > middle)
      greaters.add(last);

    List<Integer> result = new ArrayList<>();
    result.addAll(lessers);
    result.add(middle);
    result.addAll(greaters);
    return result;
  }
```

Now, don't freak out. Let's walk through this together.

First, we extract the three values into the three named variables: `first`, `middle`, and `last`. We do this for convenience because we don't want a bunch of `list.get(x)` calls littering up the code.

Next, we create a new list for the elements that are less than the `middle` and another for the elements that are greater than the `middle`. Note that we are assuming the `middle` is unique in the list.

Then, in the four subsequent `if` statements, we place the `first` and `last` elements into the appropriate lists.

Finally, we construct the `result` list by placing the `lessers`, the `middle`, and then the `greaters` into it.

Now you may not like this code. I don't much care for it either. But it works. The tests pass.

And the next two tests pass as well:

```
assertEquals(asList(1, 2, 3), sort(asList(1, 3, 2)));
assertEquals(asList(1, 2, 3), sort(asList(3, 2, 1)));
```

So far, we've tried four of the six possible cases for a list of three unique elements. If we had tried the other two, `[2,3,1]` and `[3,1,2]`, as we should have, both of them would have failed.

But due either to impatience or oversight, we move on to test a lists with four elements:

```
assertEquals(asList(1, 2, 3, 4), sort(asList(1, 2, 3, 4)));
```

This fails, of course, because the current solution assumes that the list has no more than three elements. And, of course, our simplification of `first`, `middle`, and `last` breaks down with four elements. This may make you wonder why we chose the `middle` as element 1. Why couldn't it be element 0?

So, let's comment out that last test and change middle to element 0:

```
int first = list.get(1);
int middle = list.get(0);
int last = list.get(2);
```

Surprise—the `[1,3,2]` test fails. Can you see why? If `middle` is 1, then the 3 and 2 get added to the `greaters` list in the wrong order.

Now, it just so happens that our solution already knows how to sort a list with two elements in it. And `greaters` is such a list, so we can make this pass by calling `sort` on the `greaters` list:

```
List<Integer> result = new ArrayList<>();
result.addAll(lessers);
result.add(middle);
result.addAll(sort(greaters));
return result;
```

That caused the [1,3,2] test to pass but failed the [3,2,1] test because the lessers list was out of order. But that's a pretty easy fix:

```
List<Integer> result = new ArrayList<>();
result.addAll(sort(lessers));
result.add(middle);
result.addAll(sort(greaters));
return result;
```

So, yeah, we should have tried the two remaining cases of three elements before going on to the four-element list.

> *Rule 7: Exhaust the current simpler case before testing the next more complex case.*

Anyway, now we need to get that four-element list to pass. So, we uncommented the test and saw it fail (not shown).

The algorithm we currently have for sorting the three-element list can be generalized, especially now that the middle variable is the first element of the list. All we have to do to build the lessers and greaters lists is apply filters:

```
else {
  int middle = list.get(0);
  List<Integer> lessers =
    list.stream().filter(x -> x<middle).collect(toList());
  List<Integer> greaters =
    list.stream().filter(x -> x>middle).collect(toList());

  List<Integer> result = new ArrayList<>();
  result.addAll(sort(lessers));
  result.add(middle);
```

```
    result.addAll(sort(greaters));
    return result;
  }
```

It should come as no surprise that this passes and also passes the next two tests:

```
assertEquals(asList(1, 2, 3, 4), sort(asList(2, 1, 3, 4)));
assertEquals(asList(1, 2, 3, 4), sort(asList(4, 3, 2, 1)));
```

But now you may be wondering about that `middle`. What if the `middle` element was not unique in the list? Let's try that:

```
assertEquals(asList(1, 1, 2, 3), sort(asList(1, 3, 1, 2)));
```

Yeah, that fails. That just means we should stop treating the `middle` as something special:

```
else {
  int middle = list.get(0);
  List<Integer> middles =
    list.stream().filter(x -> x == middle).collect(toList());
  List<Integer> lessers =
    list.stream().filter(x -> x<middle).collect(toList());
  List<Integer> greaters =
    list.stream().filter(x -> x>middle).collect(toList());

  List<Integer> result = new ArrayList<>();
  result.addAll(sort(lessers));
  result.addAll(middles);
  result.addAll(sort(greaters));
  return result;
}
```

This passes. However, look up at that `else`. Remember what's above it? Here, I'll show you:

```
if (list.size() <= 1)
  return list;
else if (list.size() == 2){
  int first = list.get(0);
  int second = list.get(1);
  if (first > second)
    return asList(second, first);
  else
    return asList(first, second);
}
```

Is that `==2` case really necessary anymore? No. Removing it still passes all tests.

Okay, so what about that first `if` statement? Is that still necessary? Actually, it can be changed to something better. And, in fact, let me just show you the final algorithm:

```
private List<Integer> sort(List<Integer> list) {
  List<Integer> result = new ArrayList<>();

  if (list.size() == 0)
    return result;
  else {
    int middle = list.get(0);
    List<Integer> middles =
      list.stream().filter(x -> x == middle).collect(toList());
    List<Integer> lessers =
      list.stream().filter(x -> x < middle).collect(toList());
    List<Integer> greaters =
      list.stream().filter(x -> x > middle).collect(toList());

    result.addAll(sort(lessers));
    result.addAll(middles);
    result.addAll(sort(greaters));
    return result;
  }
}
```

This algorithm has a name. It's called *quick sort*. It is one of the best sorting algorithms known.

How much better it is? This algorithm, on my laptop, can sort an array of 1 million random integers between zero and a million in 1.5 seconds. The bubble sort from the previous section will sort the same list in about six months. So . . . better.

And this leads us to a disturbing observation. There were two different solutions to sorting a list with two elements out of order. One solution led us directly to a bubble sort, the other solution led us directly to a quick sort.

This means that identifying forks in the road, and choosing the right path, can sometimes be pretty important. In this case, one path led us to a pretty poor algorithm, and the other led us to a very good algorithm.

Can we identify these forks and determine which path to choose? Perhaps. But that's a topic for a more advanced chapter.

Getting Stuck（卡壳）

At this point, I think you've seen enough videos to get a good idea about the rhythm of TDD. From now on, we'll forego the videos and just rely on the text of the chapter.

It often happens, to TDD novices, that they find themselves in a pickle. They write a perfectly good test and then discover that the only way to make that test pass is to implement the entire algorithm at once. I call this "getting stuck."

The solution to getting stuck is to delete the last test you wrote and find a simpler test to pass.

> *Rule 8: If you must implement too much to get the current test to pass, delete that test and write a simpler test that you can more easily pass.*

I often use the following exercise in my classroom sessions to get people stuck. It's pretty reliable. Well over half the people who try it find themselves stuck and also find it difficult to back out.

The problem is the good-old word-wrap problem: Given a string of text without any line breaks, insert appropriate line breaks so that the text will fit in a column `N` characters wide. Break at words if at all possible.

Students are supposed to write the following function:

```
Wrapper.wrap(String s, int w);
```

Let's suppose that the input string is the Gettysburg Address:

```
"Four score and seven years ago our fathers brought forth upon this
continent a new nation conceived in liberty and dedicated to the
proposition that all men are created equal"
```

Now, if the desired width is 30, then the output should be

```
====:====:====:====:====:====:
Four score and seven years ago
Our fathers brought forth upon
This continent a new nation
Conceived in liberty and
Dedicated to the proposition
That all men are created equal
====:====:====:====:====:====:
```

How would you write this algorithm test-first? We might begin as usual with this failing test:

```
public class WrapTest {
  @Test
  public void testWrap() throws Exception {
    assertEquals("Four", wrap("Four", 7));
  }
```

```
  private String wrap(String s, int w) {
    return null;
  }
}
```

How many TDD rules did we break with this test? Can you name them? Let's proceed anyway. It's easy to make this test pass:

```
private String wrap(String s, int w) {
  return "Four";
}
```

The next test seems pretty obvious:

```
assertEquals("Four\nscore", wrap("Four score", 7));
```

And the code that makes that pass is pretty obvious too:

```
private String wrap(String s, int w) {
  return s.replace(" ", "\n");
}
```

Just replace all spaces with line ends. Perfect. Before we continue, let's clean this up a bit:

```
private void assertWrapped(String s, int width, String expected) {
  assertEquals(expected, wrap(s, width));
}

@Test
public void testWrap() throws Exception {
  assertWrapped("Four", 7, "Four");
  assertWrapped("Four score", 7, "Four\nscore");
}
```

That's better. Now, the next failing test. If we simply follow the Gettysburg Address along, the next failure would be

```
assertWrapped("Four score and seven years ago our", 7,
  "Four\nscore\nand\nseven\nyears\nago our");
```

That does, in fact, fail. We can tighten that failure up a bit like so:

```
assertWrapped("ago our", 7, "ago our");
```

Okay, now how do we make that pass? It looks like we need to *not* replace *all* spaces with line ends. So, which ones *do* we replace? Or, should we go ahead and replace all spaces with line ends, and then figure out which line ends to put back?

I'll let you ponder this for a while. I don't think you'll find an easy solution. And that means we're stuck. The only way to get this test to pass is to invent a very large part of the word-wrap algorithm at once.

The solution to being stuck is to delete one or more tests and replace them with simpler tests that you can pass incrementally. Let's try that:

```
@Test
public void testWrap() throws Exception {
  assertWrapped("", 1, "");
}

private String wrap(String s, int w) {
  return "";
}
```

Ah, yes, that's a truly degenerate test, isn't it? That's one of the rules we forgot earlier.

Okay, what's the next most degenerate test? How about this?

```
assertWrapped("x", 1, "x");
```

That's pretty degenerate. It's also really easy to pass:

```
private String wrap(String s, int w) {
  return s;
}
```

There's that pattern again. We get the first test to pass by returning the degenerate constant. We get the second test to pass by returning the input. Interesting. Okay, so what's the next most degenerate test?

```
assertWrapped("xx", 1, "x\nx");
```

That fails because it returns "xx". It's not hard to get it to pass, though:

```
private String wrap(String s, int w) {
  if (w >= s.length())
    return s;
  else
    return s.substring(0, w) + "\n" + s.substring(w);


}
```

That was easy. So, what's the next most degenerate test?

```
assertWrapped("xx", 2, "xx");
```

That already passes. Nice. So, the next test would be

```
assertWrapped("xxx", 1, "x\nx\nx");
```

This fails. And it implies some kind of a loop. But wait. There's an easier way:

```
private String wrap(String s, int w) {
  if (w >= s.length())
    return s;
  else
    return s.substring(0, w) + "\n" + wrap(s.substring(w), w);
}
```

We don't often think of recursion, do we? Perhaps we should think about it more often.

We've got a nice little pattern cooking in our tests, don't we? No words, not even any spaces yet. Just a string of x's with counts walking from 1 to the size of the string. So, the next test will be

```
assertWrapped("xxx", 2, "xx\nx");
```

And that one already passes. So does the next one:

```
assertWrapped("xxx", 3, "xxx");
```

There's probably no point in continuing with this pattern. It's time for some spaces:

```
assertWrapped("x x", 1, "x\nx");
```

This one fails because it returns `"x\n \nx"`. We can fix that by eliminating any prefix spaces before we do the recursive call to `wrap`.

```
return s.substring(0, w) + "\n" + wrap(s.substring(w).trim(), w);
```

This passes. And now we have a new test pattern to follow. So, the next test is

```
assertWrapped("x x", 2, "x\nx");
```

This fails because the first substring has a trailing space. We can get rid of that with another call to `trim`:

```
return s.substring(0, w).trim() + "\n" + wrap(s.substring(w).trim(), w);
```

And this passes. The next test in the sequence passes too:

```
assertWrapped("x x", 3, "x x");
```

What's next? We could try these:

```
assertWrapped("x x x", 1, "x\nx\nx");
assertWrapped("x x x", 2, "x\nx\nx");
assertWrapped("x x x", 3, "x x\nx");
assertWrapped("x x x", 4, "x x\nx");
assertWrapped("x x x", 5, "x x x");
```

They all pass. There's probably not much point in adding the fourth x.

Let's try this:

```
assertWrapped("xx xx", 1, "x\nx\nx\nx");
```

That one passes. And so do the next two tests in the sequence:

```
assertWrapped("xx xx", 2, "xx\nxx");
assertWrapped("xx xx", 3, "xx\nxx");
```

But the next test fails:

```
assertWrapped("xx xx", 4, "xx\nxx");
```

It fails because it returns `"xx  x\nx"`. And that's because it did not break on the space between the two "words." Where is that space? It's *before* the wth character. So, we need to search backwards from `w` for a space:

```
private String wrap(String s, int w) {
  if (w >= s.length())
    return s;
  else {
    int br = s.lastIndexOf(" ", w);
    if (br == -1)
      br = w;
    return s.substring(0, br).trim() + "\n" +
           wrap(s.substring(br).trim(), w);
  }
}
```

This passes. I have a feeling that we are done. But let's try a few more test cases:

```
assertWrapped("xx xx", 5, "xx xx");
assertWrapped("xx xx xx", 1, "x\nx\nx\nx\nx\nx");
assertWrapped("xx xx xx", 2, "xx\nxx\nxx");
assertWrapped("xx xx xx", 3, "xx\nxx\nxx");
assertWrapped("xx xx xx", 4, "xx\nxx\nxx");
assertWrapped("xx xx xx", 5, "xx xx\nxx");
assertWrapped("xx xx xx", 6, "xx xx\nxx");
assertWrapped("xx xx xx", 7, "xx xx\nxx");
assertWrapped("xx xx xx", 8, "xx xx xx");
```

They all pass. I think we're done. Let's try the Gettysburg Address, with a length of 15:

```
Four score and
seven years ago
our fathers
brought forth
upon this
continent a new
```

```
nation
conceived in
liberty and
dedicated to
the proposition
that all men
are created
equal
```

That looks right.

So, what did we learn? First, if you get stuck, back out of the tests that got you stuck, and start writing simpler tests. But second, when writing tests, try to apply

Rule 9: Follow a deliberate and incremental pattern that covers the test space.

Arrange, Act, Assert（安排、行动、断言）

And now for something completely different.

Many years ago, Bill Wake identified the fundamental pattern for all tests. He called it the 3A pattern, or AAA. It stands for Arrange/Act/Assert.

The first thing you do when writing a test is *arrange* the data to be tested. This is typically done in a `Setup` method, or at the very beginning of the test function. The purpose is to put the system into the state necessary to run the test.

The next thing the test does is *act*. This is when the test calls the function, or performs the action, or otherwise invokes the procedure that is the target of the test.

The last thing the test does is *assert*. This usually entails looking at the output of the act to ensure that the system is in the new desired state.

As a simple example of this pattern consider this test from the bowling game in Chapter 2:

```
@Test
public void gutterGame() throws Exception {
  rollMany(20, 0);
  assertEquals(0, g.score());
}
```

The arrange portion of this test is the creation of the `Game` in the `Setup` function and the `rollMany(20, 0)` to set up the scoring of a gutter game.

The act portion of the test is the call to `g.score()`.

The assert portion of the test is the `assertEquals`.

In the two and a half decades since I started practicing TDD, I have never found a test that did not follow this pattern.

Enter BDD（进入 BDD）

In 2003, while practicing and teaching TDD, Dan North, in collaboration with Chris Stevenson and Chris Matz, made the same discovery that Bill Wake had made. However, they used a different vocabulary: Given-When-Then (GWT).

This was the beginning of *behavior-driven development* (BDD).

At first, BDD was viewed as an improved way of writing tests. Dan and other proponents liked the vocabulary better and affixed that vocabulary into testing tools such as JBehave, and RSpec.

As an example, I can rephrase the `gutterGame` test in BDD terms as follows:

```
Given that the player rolled a game of 20 gutter balls,
When I request the score of that game,
Then the score will be zero.
```

It should be clear that some parsing would have to take place in order to translate that statement into an executable test. JBehave and RSpec provided affordances for that kind of parsing. It also ought to be clear that the TDD test and the BDD test are synonymous.

Over time, the vocabulary of BDD drew it away from testing and toward the problem of system specification. BDD proponents realized that even if the GWT statements were never executed as tests, they remained valuable as specifications of behavior.

In 2013, Liz Keogh said of BDD:

> *It's using examples to talk through how an application behaves. . . . And having conversations about those examples.*

Still, it is hard to separate BDD entirely from testing if only because the vocabulary of GWT and AAA are so obviously synonymous. If you have any doubts about that, consider the following:

- *Given* that the test data has been *Arranged*
- *When* I execute the tested *Act*
- *Then* the expected result is *Asserted*

Finite State Machines（有限状态机）

The reason I made such a big deal out of the synonymity of GWT and AAA is that there is another famous triplet that we frequently encounter in software: the transition of a finite state machine.

Consider the state/transition diagram of a simple subway turnstile (Figure 3.1).

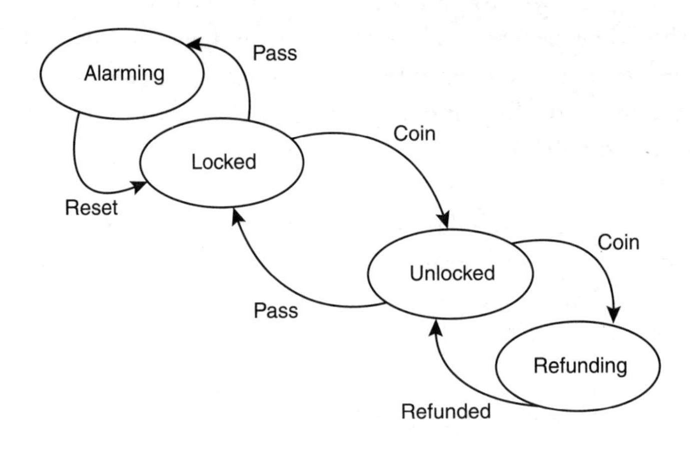

Figure 3.1 Transition/state diagram for a subway turnstile

The turnstile starts in the locked state. A coin will send it to the unlocked state. When someone passes through, the turnstile returns to the locked state. If someone passes without paying, the turnstile alarms. If someone drops two coins, the extra coin is refunded.

This diagram can be turned into a state transition table as follows:

Current State	Event	Next State
Locked	Coin	Unlocked
Locked	Pass	Alarming
Unlocked	Coin	Refunding
Unlocked	Pass	Locked
Refunding	Refunded	Unlocked
Alarming	Reset	Locked

Each row in the table is a transition from the current state to the next state triggered by the event. Each row is a triplet, just like GWT or AAA. More important, each one of those transition triplets is synonymous with a corresponding GWT or AAA triplet, as follows:

```
Given it is in the Locked state
When it gets a Coin event
Then it transitions to the Unlocked state.
```

What we can deduce from this is that *every test you write is a transition of the finite state machine that describes the behavior of the system.*

Repeat that to yourself several times. Every test is a transition of the finite state machine you are trying to create in your program.

Did you know that the program you are writing is a finite state machine? Of course it is. Every program is a finite state machine. That's because computers are nothing more than finite state machine processors.
The computer itself transitions from one finite state to the next with every instruction it executes.

Thus, the tests you write when practicing TDD and the behaviors you describe while practicing BDD are simply transitions of the finite state machine that you are attempting to create. Your test suite, if it is complete, *is that finite state machine.*

The obvious question, therefore, is how do you ensure that all the transitions you desire your state machine to handle are encoded as tests? How do you ensure that the state machine that your tests are describing is the complete state machine that your program should be implementing?

What better way than to write the transitions first, as tests, and then write the production code that implements those transitions?

BDD Again（再谈 BDD）

And don't you think it's fascinating, and perhaps even a little ironic, that the BDD folks, perhaps without realizing it, have come to the conclusion that the best way to describe the behavior of a system is by specifying it as a finite state machine?

TEST DOUBLES（测试替身）

In 2000, Steve Freeman, Tim McKinnon, and Philip Craig presented a paper[1] called "Endo-Testing: Unit Testing with Mock Objects." The influence this paper had on the software community is evidenced by the pervasiveness of the term they coined: *mock*. The term has since become a verb. Nowadays, we use *mocking* frameworks to *mock* things out.

In those early days, the notion of TDD was just beginning to permeate the software community. Most of us had never applied object-oriented design to test code. Most of us had never applied any kind of design at all to test code. This led to all kinds of problems for test authors.

Oh, we could test the simple things like the examples you saw in the previous chapters. But there was another class of problems that we just didn't know how to test. For example, how do you test the code that reacts to an input/output (IO) failure? You can't really force an IO device to fail in a unit test. Or how do you test the code that interacts with an external service? Do you have to have the external service connected for your tests? And how do you test the code that handles external service failure?

The original TDDers were Smalltalk programmers. For them, objects were what the universe was made of. So, although they were almost certainly using mock objects, they likely thought nothing of it. Indeed, when I presented the idea of a mock object in Java to one expert Smalltalker and TDDer in 1999, his response to me was: "Too much mechanism."

Nevertheless, the technique caught hold and has become a mainstay of TDD practitioners.

---

1. Steve Freeman, Tim McKinnon, and Philip Craig, "Endo-Testing: Unit Testing with Mock Objects," paper presented at eXtreme Programming and Flexible Processes in Software Engineering (XP2000), https://www2.ccs.neu.edu/research/demeter/related-work/extreme-programming/MockObjectsFinal.PDF.

But before we delve too much further into the technique itself, we need to clear up a vocabulary issue. Almost all of us use the term *mock object* incorrectly—at least in a formal sense. The mock objects we talk about nowadays are very different from the mock objects that were presented in the endo-testing paper back in 2000. So different, in fact, that a different vocabulary has been adopted to clarify the separate meanings.

In 2007, Gerard Meszaros published *xUnit Test Patterns: Refactoring Test Code*.[2] In it, he adopted the formal vocabulary that we use today. Informally, we still talk about mocks and mocking, but when we want to be precise, we use Meszaros's formal vocabulary.

Meszaros identified five kinds of objects that fall under the informal mock category: dummies, stubs, spies, mocks, and fakes. He called them all *test doubles*.

That's actually a very good name. In the movies, a stunt double stands in for an actor, and a hand double stands in for closeups of an actor's hands. A body double stands in for shots when the body, and not the face, of the actor is on screen. And that's just what a test double does. A test double stands in for another object while a test is being run.

Test doubles form a type hierarchy of sorts (Figure 3.2). Dummies are the simplest. Stubs are dummies, spies are stubs, and mocks are spies. Fakes stand alone.

---

2. Gerard Meszaros, *xUnit Test Patterns: Refactoring Test Code* (Addison-Wesley, 2007).

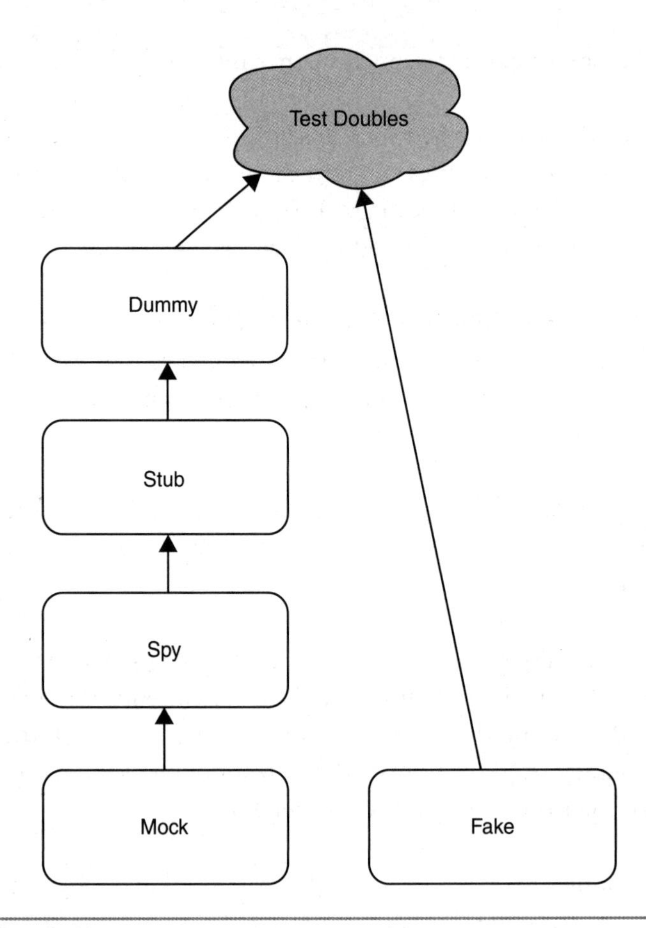

Figure 3.2 Test doubles

The mechanism that all test doubles use (and which my Smalltalker friend thought was "too much") is simply polymorphism. For example, if you want to test the code that manages an external service, then you isolate that external service behind a polymorphic interface, and then you create an implementation of that interface that stands in for the service. That implementation is the test double.

But perhaps the best way to explain all this is to demonstrate it.

Dummy

Test doubles generally start with an interface—an abstract class with no implemented methods. For example, we could start with the Authenticator interface:

```
public interface Authenticator {
  public Boolean authenticate(String username, String password);
}
```

The intent of this interface is to provide our application with a way to authenticate users by using usernames and passwords. The authenticate function returns true if the user is authentic and false if not.

Now let's suppose that we want to test that a LoginDialog can be cancelled by clicking the close icon before the user enters a username and password. That test might look like this:

```
@Test
public void whenClosed_loginIsCancelled() throws Exception {
  Authenticator authenticator = new ???;
  LoginDialog dialog = new LoginDialog(authenticator);
  dialog.show();
  boolean success = dialog.sendEvent(Event.CLOSE);
  assertTrue(success);
}
```

Notice that the LoginDialog class must be constructed with an Authenticator. But that Authenticator will never be called by this test, so what should we pass to the LoginDialog?

Let's further assume that the RealAuthenticator is an expensive object to create because it requires a DatabaseConnection passed in to its constructor. And let's say that the DatabaseConnection class has a constructor that requires valid UIDs for a databaseUser and databaseAuthCode. (I'm sure you've seen situations like this.)

```
public class RealAuthenticator implements Authenticator {
  public RealAuthenticator(DatabaseConnection connection) {
    //...
  }

  //...

}

public class DatabaseConnection {
  public DatabaseConnection(UID databaseUser, UID databaseAuthCode) {
    //...
  }
}
```

To use the RealAuthenticator in our test, we'd have to do something horrible, like this:

```
@Test
public void whenClosed_loginIsCancelled() throws Exception {
  UID dbUser = SecretCodes.databaseUserUID;
  UID dbAuth = SecretCodes.databaseAuthCode;
  DatabaseConnection connection =
    new DatabaseConnection(dbUser, dbAuth);
  Authenticator authenticator = new RealAuthenticator(connection);
  LoginDialog dialog = new LoginDialog(authenticator);
  dialog.show();
  boolean success = dialog.sendEvent(Event.CLOSE);
  assertTrue(success);
}
```

That's a terrible load of cruft to put into our test just so that we can create an Authenticator that will never be used. It also adds two dependencies to our test, which the test does not need. Those dependencies could break our test at either compile or load time. We don't need the mess or the headache.

Rule 10: Don't include things in your tests that your tests don't need.

So, what we do instead is use a dummy (Figure 3.3).

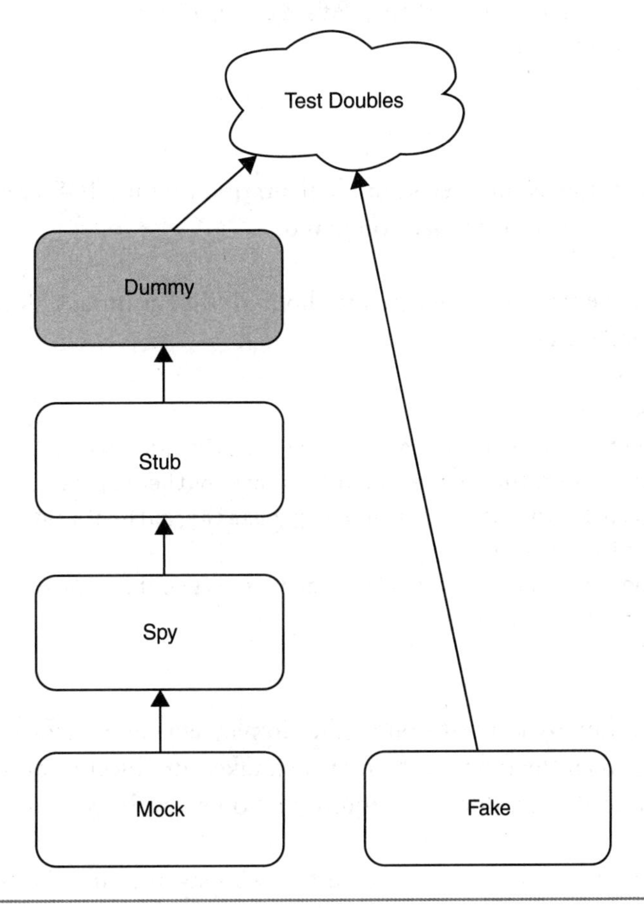

Figure 3.3 Dummy

A dummy is an implementation that does *nothing*. Every method of the interface is implemented to do *nothing*. If a method returns a value, then the value returned by the dummy will be as close as possible to `null` or zero.

In our example, the `AuthenticatorDummy` would look like this:

```
public class AuthenticatorDummy implements Authenticator {
  public Boolean authenticate(String username, String password) {
    return null;
  }
}
```

In fact, this is the precise implementation that my IDE creates when I invoke the *Implement Interface* command.

Now the test can be written without all that cruft and all those nasty dependencies:

```
@Test
public void whenClosed_loginIsCancelled() throws Exception {
  Authenticator authenticator = new AuthenticatorDummy();
  LoginDialog dialog = new LoginDialog(authenticator);
  dialog.show();
  boolean success = dialog.sendEvent(Event.CLOSE);
  assertTrue(success);
}
```

So, a dummy is a test double that implements an interface to do nothing. It is used when the function being tested takes an object as an argument, but the *logic* of the test does not require that object to be present.

I don't use dummies very often for two reasons. First, I don't like having functions with code pathways that don't use the arguments of that function. Second, I don't like objects that have chains of dependencies such as `LoginDialog->Authenticator->DatabaseConnection->UID`. Chains like that always cause trouble down the road.

Still, there are times when these problems cannot be avoided, and in those situations, I much prefer to use a dummy rather than fight with complicated objects from the application.

STUB

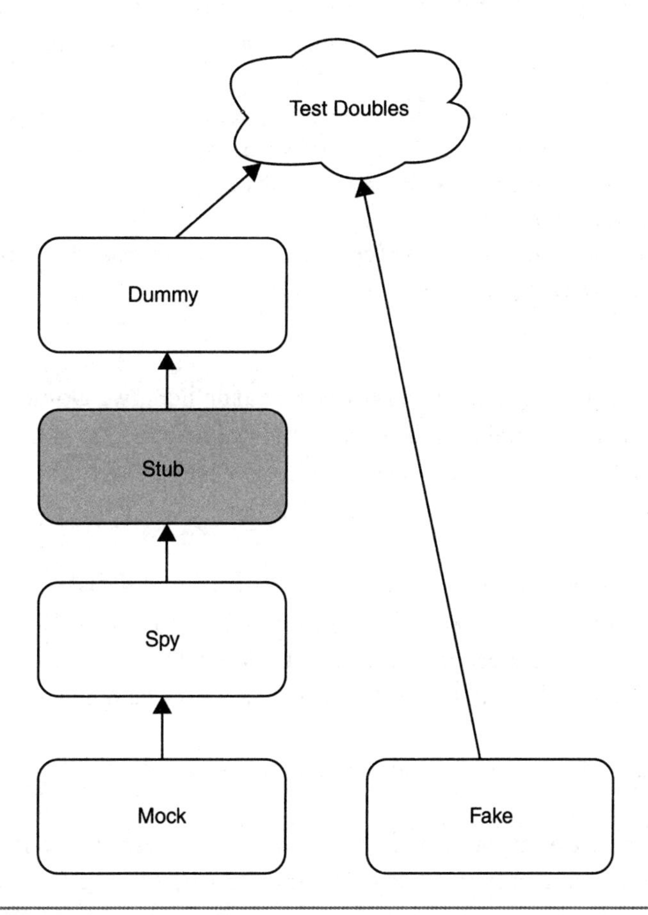

Figure 3.4 Stub

As Figure 3.4 shows, a *stub* is a dummy; it is implemented to do nothing. However, instead of returning zero or `null`, the functions of a stub return values that drive the function being tested through the pathways that the test wants executed.

Let's imagine the following test that ensures that a login attempt will fail if the Authenticator rejects the username and password:

```
public void whenAuthorizerRejects_loginFails() throws Exception {
  Authenticator authenticator = new ?;
  LoginDialog dialog = new LoginDialog(authenticator);
  dialog.show();
  boolean success = dialog.submit("bad username", "bad password");
  assertFalse(success);
}
```

If we were to use the RealAuthenticator here, we would still have the problem of initializing it with all the cruft of the DatabaseConnection and the UIDs. But we'd also have another problem. What username and password should we use?

If we know the contents of the authorization database, then we could select a username and password that we know is not present. But that's a horrible thing to do because it creates a data dependency between our tests and production data. If that production data ever changes, it could break our test.

Rule 11: Don't use production data in your tests.

What we do instead is create a stub. For this test, we need a RejectingAuthenticator that simply returns false from the authorize method:

```
public class RejectingAuthenticator implements Authenticator {
  public Boolean authenticate(String username, String password) {
    return false;
  }
}
```

And now we can simply use that stub in our test:

```
public void whenAuthorizerRejects_loginFails() throws Exception {
  Authenticator authenticator = new RejectingAuthenticator();
  LoginDialog dialog = new LoginDialog(authenticator);
  dialog.show();
  boolean success = dialog.submit("bad username", "bad password");
  assertFalse(success);
}
```

We expect that the `submit` method of the `LoginDialog` will call the `authorize` function, and we know that the `authorize` function will return `false`, so we know what pathway the code in the `LoginDialog.submit` `method` will take; and that is precisely the path we are testing.

If we want to test that login succeeds when the `authorizer` accepts the `username` and `password`, we can play the same game with a different stub:

```
public class PromiscuousAuthenticator implements Authenticator {
  public Boolean authenticate(String username, String password) {
    return true;
  }
}
@Test
public void whenAuthorizerAccepts_loginSucceeds() throws Exception {
  Authenticator authenticator = new PromiscuousAuthenticator();
  LoginDialog dialog = new LoginDialog(authenticator);
  dialog.show();
  boolean success = dialog.submit("good username", "good password");
  assertTrue(success);
}
```

So, a stub is a dummy that returns test-specific values in order to drive the system under test through the pathways being tested.

Spy

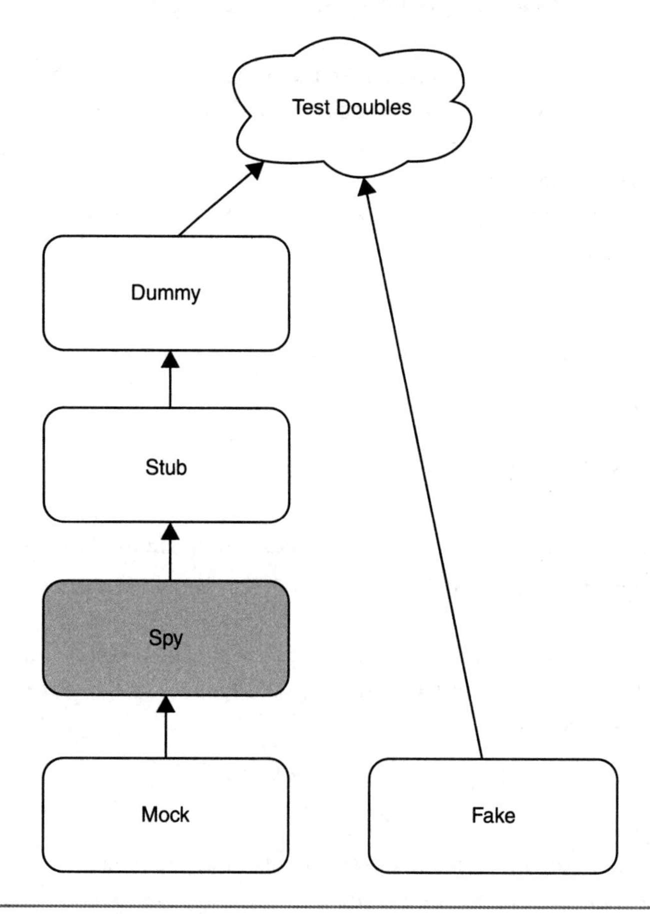

Figure 3.5 Spy

A *spy* (Figure 3.5) is a stub. It returns test-specific values in order to drive the system under test through desired pathways. However, a spy remembers what was done to it and allows the test to ask about it.

The best way to explain that is with an example:

```
public class AuthenticatorSpy implements Authenticator {
  private int count = 0;
  private boolean result = false;
  private String lastUsername = "";
  private String lastPassword = "";

  public Boolean authenticate(String username, String password) {
    count++;
    lastPassword = password;
    lastUsername = username;
    return result;
  }

  public void setResult(boolean result) {this.result = result;}
  public int getCount() {return count;}
  public String getLastUsername() {return lastUsername;}
  public String getLastPassword() {return lastPassword;}
}
```

Note that the `authenticate` method keeps track of the number of times it was called and the last `username` and `password` that it was called with. Notice also that it provides accessors for these values. It is that behavior and those accessors that make this class a spy.

Notice also that the `authenticate` method returns `result`, which can be set by the `setResult` method. That makes this spy a programmable stub.

Here's a test that might use that spy:

```
@Test
public void loginDialog_correctlyInvokesAuthenticator() throws
Exception {
  AuthenticatorSpy spy = new AuthenticatorSpy();
  LoginDialog dialog = new LoginDialog(spy);
  spy.setResult(true);
  dialog.show();
  boolean success = dialog.submit("user", "pw");
  assertTrue(success);
  assertEquals(1, spy.getCount());
  assertEquals("user", spy.getLastUsername());
  assertEquals("pw", spy.getLastPassword());
}
```

The name of the test tells us a lot. This test makes sure that the `LoginDialog` correctly invokes the `Authenticator`. It does this by making sure that the `authenticate` method is called only once and that the arguments were the arguments that were passed into `submit`.

A spy can be as simple as a single Boolean that is set when a particular method is called. Or a spy can be a relatively complex object that maintains a history of every call and every argument passed to every call.

Spies are useful as a way to ensure that the algorithm being tested behaves correctly. Spies are dangerous because they couple the tests to the *implementation* of the function being tested. We'll have more to say about this later.

Mock

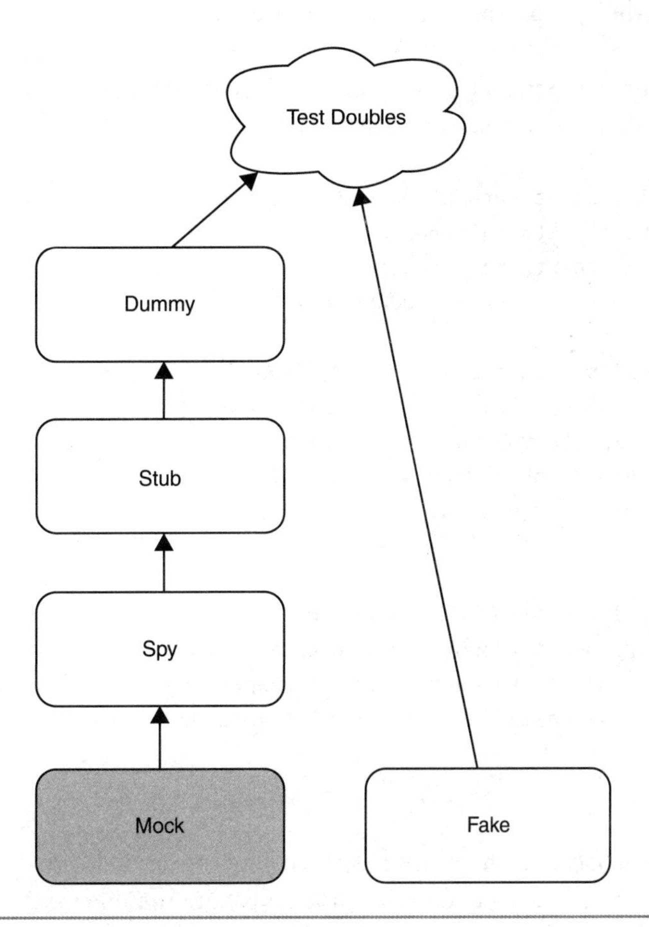

Figure 3.6 The mock object

Now, at last, we come to the true mock object (Figure 3.6). This is the mock that Mackinnon, Freeman, and Craig described in their endo-testing paper.

A *mock* is a spy. It returns test-specific values in order to drive the system under test through desired pathways, and it remembers what was done to it. However, a mock also knows what to expect and will pass or fail the test on the basis of those expectations.

In other words, the test assertions are written into the mock.

Again, an explanation in code is worth all the words I can write about this, so let's build an AuthenticatorMock:

```
public class AuthenticatorMock extends AuthenticatorSpy{
  private String expectedUsername;
  private String expectedPassword;
  private int expectedCount;

  public AuthenticatorMock(String username, String password,
                           int count) {
    expectedUsername = username;
    expectedPassword = password;
    expectedCount = count;
  }

  public boolean validate() {
    return getCount() == expectedCount &&
      getLastPassword().equals(expectedPassword) &&
      getLastPassword().equals(expectedUsername);
  }
}
```

As you can see, the mock has three expectation fields that are set by the constructor. This makes this mock a programmable mock. Notice also that the AuthenticatorMock derives from the AuthenticatorSpy. We reuse all that spy code in the mock.

The validate function of the mock does the final comparison. If the count, lastPassword, and lastUsername collected by the spy match the expectations set into the mock, then validate returns true.

Now the test that uses this mock should make perfect sense:

```
@Test
public void loginDialogCallToAuthenticator_validated() throws
Exception {
  AuthenticatorMock mock = new AuthenticatorMock("Bob", "xyzzy", 1);
  LoginDialog dialog = new LoginDialog(mock);
  mock.setResult(true);
  dialog.show();
  boolean success = dialog.submit("Bob", "xyzzy");
  assertTrue(success);
  assertTrue(mock.validate());
}
```

We create the mock with the appropriate expectations. The `username` should be `"Bob"`, the `password` should be `"xyzzy"`, and the number of times the `authenticate` method is called should be `1`.

Next, we create the `LoginDialog` with the mock, which is also an `Authenticator`. We set the mock to return success. We show the dialog. We submit the login request with `"Bob"` and `"xyzzy"`. We ensure that the login succeeded. And then we assert the mock's expectations were met.

That's a mock object. You can imagine that mock objects can get very complicated. For example, you might expect function `f` to be called three times with three different sets of arguments, returning three different values. You might also expect function `g` to be called once between the first two calls to `f`. Would you even dare to write that mock without unit tests for the mock itself?

I don't much care for mocks. They couple the spy behavior to the test assertions. That bothers me. I think a test should be very straightforward about what it asserts and should not defer those assertions to some other, deeper mechanism. But that's just me.

Fake

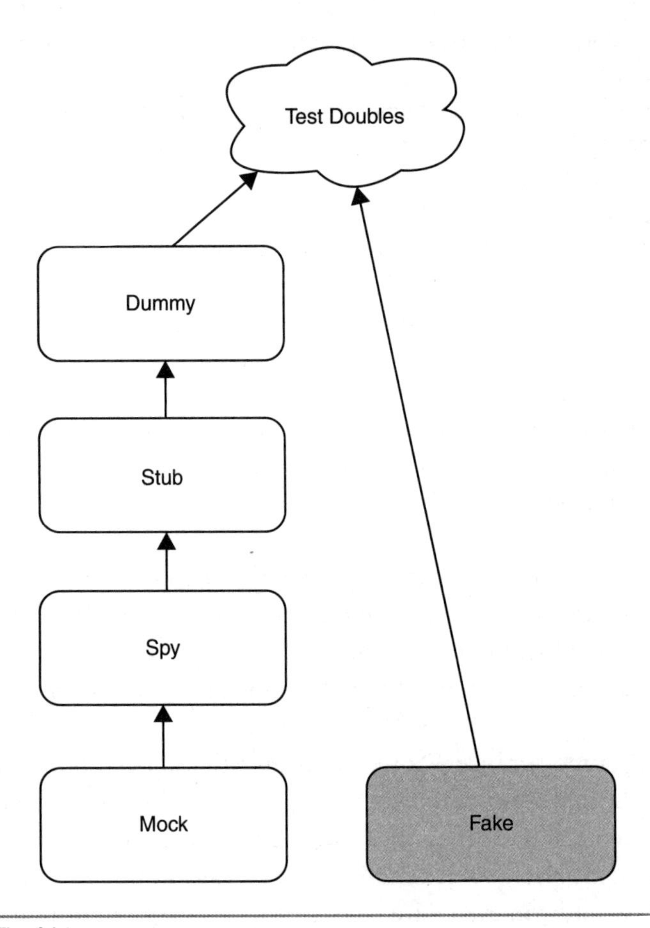

Figure 3.7 The fake

Finally, we can deal with the last of the test doubles: the *fake* (Figure 3.7). A fake is not a dummy, not a stub, not a spy, and not a mock. A fake is a different kind of test double entirely. A fake is a simulator.

Long ago, in the late 1970s, I worked for a company that built a system that was installed into telephone company facilities. This system tested telephone lines. There was a central computer at the service center that communicated over modem links to computers we installed in the switching offices. The computer in the service center was called the SAC (service area computer), and the computer in the switching office was called the COLT (central office line tester).

The COLT interfaced to the switching hardware and could create an electrical connection between any of the phone lines emanating from that switching office and the measurement hardware that it controlled. The COLT would then measure the electronic characteristics of the phone line and report the raw results back to the SAC.

The SAC did all the analysis on those raw results in order to determine whether there was a fault and, if so, where that fault was located.

How did we test that system?

We built a fake. The fake we built was a COLT whose switching interface was replaced with a simulator. That simulator would pretend to dial up phone lines and pretend to measure them. Then it would report back canned raw results based on the phone number it was asked to test.

The fake allowed us to test the SAC communication, control, and analysis software without having to install an actual COLT in a real telephone company switching office or even having to install real switching hardware and "real" phone lines.

Today, a fake is a test double that implements some kind of rudimentary business rules so that the tests that use that fake can select how the fake behaves. But perhaps an example would be the best explanation:

```
@Test
public void badPasswordAttempt_loginFails() throws Exception {
  Authenticator authenticator = new FakeAuthenticator();
  LoginDialog dialog = new LoginDialog(authenticator);
  dialog.show();
  boolean success = dialog.submit("user", "bad password");
  assertFalse(success);
}
```

```
@Test
public void goodPasswordAttempt_loginSucceeds() throws Exception {
  Authenticator authenticator = new FakeAuthenticator();
  LoginDialog dialog = new LoginDialog(authenticator);
  dialog.show();
  boolean success = dialog.submit("user", "good password");
  assertTrue(success);
}
```

These two tests use the same `FakeAuthorizer` but pass it a different password. The tests expect that `bad password` will fail the login attempt and that `good password` will succeed.

The code for `FakeAuthenticator` should be easy to envision:

```
public class FakeAuthenticator implements Authenticator {
  public Boolean authenticate(String username, String password)
  {
    return (username.equals("user") &&
            password.equals("good password"));
  }
}
```

The problem with fakes is that as the application grows, there will always be more conditions to test. Consequently, the fakes tend to grow with each new tested condition. Eventually, they can get so large and complex that they need tests of their own.

I seldom write fakes because I don't trust them not to grow.

The TDD Uncertainty Principle（TDD 不确定性原理）

To mock or not to mock, that is the question. Actually, no. The question really is *when* to mock.

There are two schools of thought about this: the London school and the Chicago school, which are addressed at the end of this chapter. But before we get into that, we need to define why this is an issue in the first place. It is an issue because of the *TDD uncertainty principle*.

To help us understand this, I want you to indulge me in a playful bit of "going to extremes." What follows is not something you would ever really do, but it illustrates quite nicely the point I'm trying to make.

Imagine that we want to use TDD to write a function that calculates the trigonometric sine of an angle represented in radians. What's the first test?

Remember, we like to start with the most degenerate case. Let's test that we can calculate the sine of zero:

```
public class SineTest {
  private static final double EPSILON = 0.0001;
  @Test
  public void sines() throws Exception {
    assertEquals(0, sin(0), EPSILON);
  }

  double sin(double radians) {
    return 0.0;
  }
}
```

Now, if you are thinking ahead, this should already bother you. This test does not constrain anything other than the value of `sin(0)`.

What do I mean by that? I mean that most of the functions we write using TDD are so constrained by the growing set of tests that there comes a point where the function will pass any other test we can pose. We saw that in the prime factors example and the bowling game example. Each test narrowed the possible solution until finally the solution was known.

But here, the `sin(r)` function does not look to behave that way. The test for `sin(0) == 0` is correct, but it does not appear to constrain the solution beyond that one point.

This becomes much more evident when we try the next test. What should that test be? Why not try sin(π)?

```
public class SineTest {
  private static final double EPSILON = 0.0001;
  @Test
  public void sines() throws Exception {
    assertEquals(0, sin(0), EPSILON);
    assertEquals(0, sin(Math.PI), EPSILON);
  }

  double sin(double radians) {
    return 0.0;
  }
}
```

Once again, we have that feeling of being unconstrained. This test doesn't seem to add anything to the solution. It gives us no hint of how to solve the problem, so let's try π/2:

```
public class SineTest {
  private static final double EPSILON = 0.0001;
  @Test
  public void sines() throws Exception {
    assertEquals(0, sin(0), EPSILON);
    assertEquals(0, sin(Math.PI), EPSILON);
    assertEquals(1, sin(Math.PI/2), EPSILON);
  }
  double sin(double radians) {
    return 0.0;
  }
}
```

This fails. How can we make it pass? Again, the test gives us no hint about how to pass it. We could try to put some horrible `if` statement in, but that will just lead to more and more `if` statements.

At this point, you might think that the best approach would be to look up the Taylor series for sine and just implement that.

$$x-\frac{x^3}{3!}+\frac{x^5}{5!}-\frac{x^7}{7!}+\cdots$$

That shouldn't be too hard:

```
public class SineTest {
  private static final double EPSILON = 0.0001;
  @Test
  public void sines() throws Exception {
    assertEquals(0, sin(0), EPSILON);
    assertEquals(0, sin(Math.PI), EPSILON);
    assertEquals(1, sin(Math.PI/2), EPSILON);
  }

  double sin(double radians) {
    double r2 = radians * radians;
    double r3 = r2*radians;
    double r5 = r3 * r2;
    double r7 = r5 * r2;
    double r9 = r7 * r2;
    double r11 = r9 * r2;
    double r13 = r11 * r2;
    return (radians - r3/6 + r5/120 - r7/5040 + r9/362880 -
            r11/39916800.0 + r13/6227020800.0);
  }
}
```

This passes, but it's pretty ugly. Still, we ought to be able to calculate a few other sines this way:

```
public void sines() throws Exception {
  assertEquals(0, sin(0), EPSILON);
  assertEquals(0, sin(Math.PI), EPSILON);
  assertEquals(1, sin(Math.PI/2), EPSILON);
  assertEquals(0.8660, sin(Math.PI/3), EPSILON);
  assertEquals(0.7071, sin(Math.PI/4), EPSILON);
  assertEquals(0.5877, sin(Math.PI/5), EPSILON);
}
```

Yes, this passes. But this solution is ugly because it is limited in precision. We should take the Taylor series out to enough terms that it converges to the limit of our precision. (Note the change to the ESPILON constant.)

```
public class SineTest {
  private static final double EPSILON = 0.000000001;
  @Test
  public void sines() throws Exception {
    assertEquals(0, sin(0), EPSILON);
    assertEquals(0, sin(Math.PI), EPSILON);
    assertEquals(1, sin(Math.PI/2), EPSILON);
    assertEquals(0.8660254038, sin(Math.PI/3), EPSILON);
    assertEquals(0.7071067812, sin(Math.PI/4), EPSILON);
    assertEquals(0.5877852523, sin(Math.PI/5), EPSILON);
  }

  double sin(double radians) {
    double result = radians;
    double lastResult = 2;
    double m1 = -1;
    double sign = 1;
    double power = radians;
    double fac = 1;
    double r2 = radians * radians;
    int n = 1;
    while (!close(result, lastResult)) {
      lastResult = result;
      power *= r2;
      fac *= (n+1) * (n+2);
      n += 2;
      sign *= m1;
      double term = sign * power / fac;
      result += term;
    }

    return result;
  }

  boolean close(double a, double b) {
    return Math.abs(a - b) < .0000001;
```

```
  }
}
```

Okay, now we're cooking with gas. But wait? What happened to the TDD? And how do we know that this algorithm is really working right? I mean, that's a lot of code. How can we tell if that code is right?

I suppose we could test a few more values. And, yikes, these tests are getting unwieldy. Let's refactor a bit too:

```
private void checkSin(double radians, double sin) {
  assertEquals(sin, sin(radians), EPSILON);
}

@Test
public void sines() throws Exception {
  checkSin(0, 0);
  checkSin(PI, 0);
  checkSin(PI/2, 1);
  checkSin(PI/3, 0.8660254038);
  checkSin(PI/4, 0.7071067812);
  checkSin(PI/5, 0.5877852523);

  checkSin(3* PI/2, -1);
}
```

Okay, that passes. Let's try a couple more:

```
checkSin(2*PI, 0);
checkSin(3*PI, 0);
```

Ah, 2π works, but 3π does not. It's close, though: `4.6130E-9`. We could probably fix that by bumping up the limit of our comparison in the `close()` function, but that seems like cheating and probably wouldn't work for 100π or 1,000π. A better solution would be to reduce the angle to keep it between 0 and 2π.

```
double sin(double radians) {
  radians %= 2*PI;
  double result = radians;
```

Yup. That works. Now what about negative numbers?

```
checkSin(-PI, 0);
checkSin(-PI/2, -1);
checkSin(-3*PI/2, 1);
checkSin(-1000*PI, 0);
```

Yeah, they all work. Okay, what about big numbers that aren't perfect multiples of 2π?

```
checkSin(1000*PI + PI/3, sin(PI/3));
```

Sigh. That works too. Is there anything else to try? Are there any values that might fail?

Ouch! I don't know.

The TDD Uncertainty Principle

Welcome to the first half of the TDD uncertainty principle. No matter how many values we try, we're going to be left with this nagging uncertainty that we missed something—that some input value will not produce the right output value.

Most functions don't leave you hanging like this. Most functions have the nice quality that when you've written the last test, you *know* they work. But then there are these annoying functions that leave you wondering whether there is some value that will fail.

The only way to solve that problem with the kinds of tests we've written is to try every single possible value. And since `double` numbers have 64 bits, that means we'd need to write just under 2×10^{19} tests. That's more than I'd like to write.

So, what *do* we trust about this function? Do we trust that the Taylor series calculates the sine of a given angle in radians? Yes, we saw the mathematical proof for that, so we're quite certain that the Taylor series will converge on the right value.

How can we turn the trust in the Taylor series into a set of tests that will prove we are executing that Taylor series correctly?

I suppose we could inspect each of the terms of the Taylor expansion. For example, when calculating sin(π), the terms of the Taylor series are 3.141592653589793, −2.0261201264601763, 0.5240439134171688, −0.07522061590362306, 0.006925270707505149, −4.4516023820919976E-4, 2.114256755841263E-5, −7.727858894175775E-7, 2.2419510729973346E-8.

I don't see why that kind of test is any better than the tests we already have. Those values apply to only one particular test, and they tell us nothing about whether those terms would be correct for any other value.

No, we want something different. We want something dispositive. Something that *proves* that the algorithm we are using does, in fact, execute the Taylor series appropriately.

Okay, so what is that Taylor series? It is the infinite and alternating sum of the odd powers of x divided by the odd factorials:

$$\sum_{n=1}^{\infty}(-1)^{(n-1)}\frac{x^{2n-1}}{(2n-1)!}$$

Or, in other words,

$$x-\frac{x^3}{3!}+\frac{x^5}{5!}-\frac{x^7}{7!}+\frac{x^9}{9!}-\cdots$$

How does this help us? Well, what if we had a spy that told us how the terms of the Taylor series were being calculated, it would let us write a test like this:

```
@Test
public void taylorTerms() throws Exception {
  SineTaylorCalculatorSpy c = new SineTaylorCalculatorSpy();
```

```
    double r = Math.random() * PI;
    for (int n = 1; n <= 10; n++) {
      c.calculateTerm(r, n);
      assertEquals(n - 1, c.getSignPower());
      assertEquals(r, c.getR(), EPSILON);
      assertEquals(2 * n - 1, c.getRPower());
      assertEquals(2 * n - 1, c.getFac());
    }
  }
```

Using a random number for `r` and all reasonable values for `n` allows us to avoid specific values. Our interest here is that given some `r` and some `n`, the right numbers are fed into the right functions. If this test passes, we will *know* that the `sign`, the `power`, and the `factorial` calculators have been given the right inputs.

We can make this pass with the following simple code:

```
public class SineTaylorCalculator {
  public double calculateTerm(double r, int n) {
    int sign = calcSign(n-1);
    double power = calcPower(r, 2*n-1);
    double factorial = calcFactorial(2*n-1);
    return sign*power/factorial;
  }

  protected double calcFactorial(int n) {
    double fac = 1;
    for (int i=1; i<=n; i++)
      fac *= i;
    return fac;
  }

  protected double calcPower(double r, int n) {
    double power = 1;
    for (int i=0; i<n; i++)
      power *= r;
    return power;
```

```
  }

  protected int calcSign(int n) {
    int sign = 1;
    for (int i=0; i<n; i++)
      sign *= -1;
    return sign;
  }
}
```

Note that we are not testing the actual calculation functions. They are pretty simple and probably don't need testing. This is especially true in light of the other tests we are about to write.

Here's the spy:

```
package London_sine;

public class SineTaylorCalculatorSpy extends SineTaylorCalculator {
  private int fac_n;
  private double power_r;
  private int power_n;
  private int sign_n;
  public double getR() {
    return power_r;
  }

  public int getRPower() {
    return power_n;
  }

  public int getFac() {
    return fac_n;
  }

  public int getSignPower() {
    return sign_n;
  }
```

```
    protected double calcFactorial(int n) {
      fac_n = n;
      return super.calcFactorial(n);
    }

    protected double calcPower(double r, int n) {
      power_r = r;
      power_n = n;
      return super.calcPower(r, n);
    }

    protected int calcSign(int n) {
      sign_n = n;
      return super.calcSign(n);
    }

    public double calculateTerm(double r, int n) {
      return super.calculateTerm(r, n);
    }
  }
```

Given that the test passes, how hard is it to write the summing algorithm?

```
public double sin(double r) {
  double sin=0;
  for (int n=1; n<10; n++)
    sin += calculateTerm(r, n);
  return sin;
}
```

You can complain about the efficiency of the whole thing, but do you believe that it works? Does the `calculateTerm` function properly calculate the right Taylor term? Does the `sin` function properly add them together? Are 10 iterations enough? How can we test this without falling back on all those original value tests?

Here's an interesting test. All values of `sin(r)` should be between −1 and 1 (open).

```
@Test
public void testSineInRange() throws Exception {
  SineTaylorCalculator c = new SineTaylorCalculator();
  for (int i=0; i<100; i++) {
    double r = (Math.random() * 4 * PI) - (2 * PI) ;
    double sinr = c.sin(r);
    assertTrue(sinr < 1 && sinr > -1);
  }
}
```

That passed. How about this? Given this identity,

```
public double cos(double r) {
  return (sin(r+PI/2));
}
```

let's test the Pythagorean identity: sin2 + cos2 = 1.

```
@Test
public void PythagoreanIdentity() throws Exception {
  SineTaylorCalculator c = new SineTaylorCalculator();
  for (int i=0; i<100; i++) {
    double r = (Math.random() * 4 * PI) - (2 * PI) ;
    double sinr = c.sin(r);
    double cosr = c.cos(r);
    assertEquals(1.0, sinr * sinr + cosr * cosr, 0.00001);
  }
}
```

Hmmm. That actually failed until we raised the number of terms to 20, which is, of course, an absurdly high number. But, like I said, this is an extreme exercise.

Given these tests, how confident are we that we are calculating the sine properly? I don't know about you, but I'm pretty confident. I know the terms are being fed the right numbers. I can eyeball the simple calculators, and the `sin` function seems to have the properties of a sine.

Oh, bother, let's just do some value tests for the hell of it:

```
@Test
public void sineValues() throws Exception {
    checkSin(0, 0);
    checkSin(PI, 0);
    checkSin(PI/2, 1);
    checkSin(PI/3, 0.8660254038);
    checkSin(PI/4, 0.7071067812);
    checkSin(PI/5, 0.5877852523);
}
```

Yeah, it all works. Great. I've solved my confidence problem. I am no longer uncertain that we're properly calculating sines. Thank goodness for that spy!

The TDD Uncertainty Principle (Again)

But wait. Did you know that there is a better algorithm for calculating sines? It's called CORDIC. No, I'm not going to describe it to you here. It's way beyond the scope of this chapter. But let's just say that we wanted to change our function to use that CORDIC algorithm.

Our spy tests would break!

In fact, just look back at how much code we invested in that Taylor series algorithm. We've got two whole classes, the `SineTaylorCalculator` and the `SineTaylorCalculatorSpy`, dedicated to our old algorithm. All that code would have to go away, and a whole new testing strategy would have to be employed.

The spy tests are *fragile*. Any change to the algorithm breaks virtually all those tests, forcing us to fix or even rewrite them.

On the other hand, if we'd stayed with our original value tests, then they would continue to pass with the new CORDIC algorithm. No rewriting of tests would be necessary.

Welcome to the second half of the TDD uncertainty principle. If you demand certainty from your tests, you will inevitably couple your tests to the implementation, and that will make the tests fragile.

> ***The TDD uncertainty principle:*** *To the extent you demand certainty, your tests will be inflexible. To the extent you demand flexible tests, you will have diminished certainty.*

London versus Chicago（伦敦派对决芝加哥派）

The TDD uncertainty principle might make it seem that testing is a lost cause, but that's not the case at all. The principle just puts some constraints on how beneficial our tests can be.

On the one hand, we don't want rigid, fragile tests. On the other hand, we want as much certainty as we can afford. As engineers, we have to strike the right trade-off between these two issues.

The Fragile Test Problem

Newcomers to TDD often experience the problem of fragile tests because they are not careful to design their tests well. They treat the tests as second-class citizens and break all the coupling and cohesion rules. This leads to the situation where small changes to the production code, even just a minor refactoring, cause many tests to fail and force sweeping changes in the test code.

Failed tests that necessitate significant rewriting of test code can result in initial disappointment and premature rejection of the discipline. Many young, new TDDers have walked away from the discipline simply because they failed to realize that tests have to be designed just as well as the production code.

The more you couple the tests to the production code, the more fragile your tests become; and few testing artifacts couple more tightly than spies. Spies look deep into the heart of algorithms and inextricably couple the tests to those algorithms. And since mocks are spies, this applies to mocks as well.

This is one of the reasons that I don't like mocking tools. Mocking tools often lead you to write mocks and spies, and that leads to fragile tests.

The Certainty Problem（确定性问题）

If you avoid writing spies, as I do, then you are left with value and property testing. Value tests are like the sine value tests we did earlier in this chapter. They are just the pairing of input values to output values.

Property tests are like the `testSineInRange` and `PythagoreanIdentity` tests we used earlier. They run through many appropriate input values checking for invariants. These tests can be convincing, but you are still often left with a nagging doubt.

On the other hand, these tests are so decoupled from the algorithm being employed that changing that algorithm, or even just refactoring that algorithm, cannot affect the tests.

If you are the kind of person who values certainty over flexibility, you'll likely use a lot of *spies* in your tests and you'll tolerate the inevitable fragility.

If, however, you are the kind of person who values flexibility over certainty, you'll be more like me. You'll prefer value and property tests over *spies*, and you'll tolerate the nagging uncertainty.

These two mindsets have led to two schools of TDD thought and have had a profound influence on our industry. It turns out that whether you prefer flexibility or certainty causes a dramatic change to the *process* of the design of the production code—if not to the actual design itself.

London（伦敦派）

The London school of TDD gets its name from Steve Freeman and Nat Pryce, who live in London and who wrote the book[3] on the topic. This is the school that prefers certainty over flexibility.

Note the use of the term *over*. London schoolers do not abandon flexibility. Indeed, they still value it highly. It's just that they are willing to tolerate a certain level of rigidity in order to gain more certainty.

Therefore, if you look over the tests written by London schoolers, you'll see a consistent, and relatively unconstrained, use of mocks and spies.

This attitude focuses more on algorithm than on results. To a Londoner, the results are important, but the *way in which the results are obtained* is more important. This leads to a fascinating design approach. Londoners practice *outside-in* design.

Programmers who follow the outside-in design approach start at the user interface and design their way toward the business rules, *one use case at a time*. They use mocks and spies at every boundary in order to prove that the algorithm they are using to communicate inward is working. Eventually, they get to a business rule, implement it, connect it to the database, and then turn around and test their way, with mocks and spies, back out to the user interface.

Again, this full outside-in round trip is done *one use case at a time*.

This is an extremely disciplined and orderly approach that can work very well indeed.

3. Steve Freeman and Nat Pryce, *Growing Object-Oriented Software, Guided by Tests* (Addison-Wesley, 2010).

Chicago（芝加哥派）

The Chicago school of TDD gets its name from ThoughtWorks, which was, at the time, based in Chicago. Martin Fowler was (and at the time of this writing still is) the chief scientist there. Actually, the name Chicago is a bit more mysterious than that. At one time, this school was called the Detroit school.

The Chicago school focuses on flexibility over certainty. Again, note the word *over*. Chicagoans know the value of certainty but, given the choice, prefer to make their tests more flexible. Consequently, they focus much more on results than on interactions and algorithms.

This, of course, leads to a very different design philosophy. Chicagoans tend to start with business rules and then move outward towards the user interface. This approach is often called *inside-out design*.

The Chicago design process is just as disciplined as the London process but attacks the problem in a very different order. A Chicagoan will not drive an entire use case from end to end before starting on the next. Rather, a Chicagoan might use value and property tests to implement several business rules without any user interface at all. The user interface, and the layers between it and the business rules, are implemented as and when necessary.

Chicagoans may not take the business rules all the way out to the database right away either. Rather than the one-use-case-at-a-time round trip, they are looking for synergies and duplications within the layers. Rather than sewing a thread from the inputs of a use case all the way to the output of that use case, they work in broader stripes, within the layers, starting with the business rules and moving gradually out to the user interface and database. As they explore each layer, they are hunting for design patterns and opportunities for abstraction and generality.

This approach is a less ordered than that of the London school, but it is more holistic. It tends to keep the big picture more clearly in view—in my humble opinion.

Synthesis（融合）

Although these two schools of thought exist, and although there are practitioners who tend toward one side or the other, London versus Chicago is not a war. It's not really even much of a disagreement. It is a minor point of emphasis, and little more.

Indeed, all practitioners, whether Chicagoans or Londoners, use both techniques in their work. It's just that some do a little more of one, and some do a little more of the other.

Which is right? Neither, of course. I tend more toward the Chicago side, but you may look at the London school and decide you are more comfortable there. I have no disagreement with you. Indeed, I will happily pair program with you to create a lovely synthesis.

That synthesis becomes very important when we start to consider architecture.

Architecture（架构）

The trade-off I make between the London and Chicago strategies is architectural. If you read *Clean Architecture*,[4] you know that I like to partition systems into components. I call the divisions between those components *boundaries*. My rule for boundaries is that source code dependencies must always cross a boundary pointing toward high-level policy.

This means that components that contain low-level details, like graphical user interfaces (GUIs) and databases, depend on higher-level components like business rules. High-level components do not depend on lower-level components. This is an example of the dependency inversion principle, which is the D in SOLID.

4. Robert C. Martin, *Clean Architecture: A Craftsman's Guide to Software Structure and Design* (Addison-Wesley, 2018).

When writing the lowest-level programmer tests, I will use *spies* (and rarely *mocks*) when testing across an architectural boundary. Or, to say that a different way, when testing a component, I use *spies* to mock out any collaborating components and to ensure that the component I am testing invokes the collaborators correctly. So, if my test crosses an architectural boundary, I'm a Londoner.

However, when a test does not cross such a boundary, I tend to be a Chicagoan. Within a component, I depend much more on state and property testing in order to keep the coupling, and therefore the fragility, of my tests as low as possible.

Let's look at an example. The UML diagram in Figure 3.8 shows a set of classes and the four components that contain them.

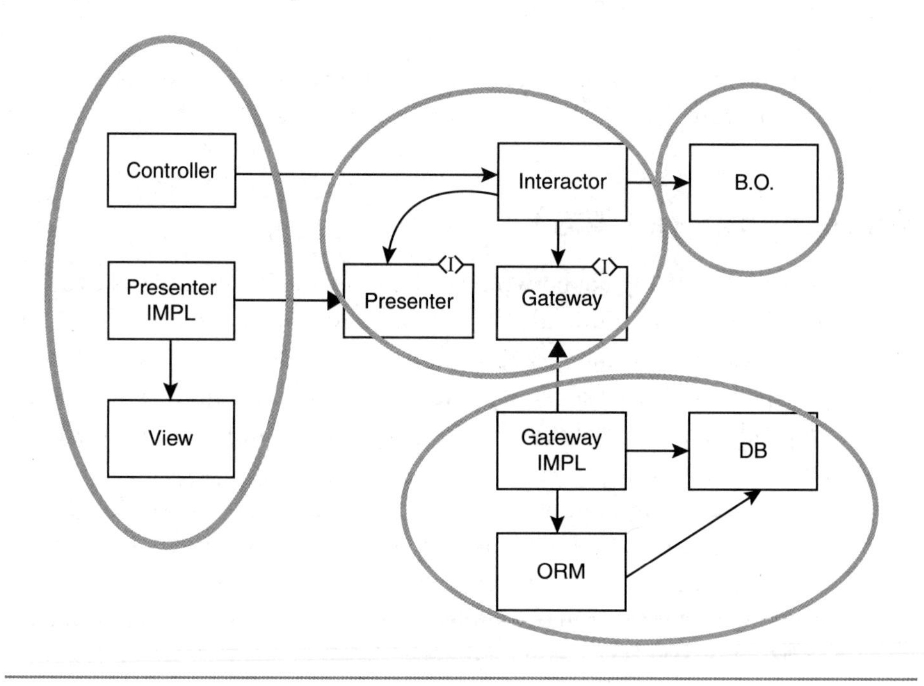

Figure 3.8 A set of classes and the four components that contain them

Note that the arrows all point from lower-level components to higher-level components. This is the *Dependency Rule* that is discussed in *Clean Architecture*. The highest-level component contains the business objects. The next level down contains the interactors and the communications interfaces. At the lowest level is the GUI and the database.

We might use some stubs when testing the business objects, but we won't need any spies or mocks because the business objects don't know about any other components.

The interactors, on the other hand, manipulate the business objects, the database, and the GUI. Our tests will likely use spies to make sure that the database and GUI are being manipulated properly. However, we will probably not use as many spies, or even stubs, between the interactors and the business objects because the functions of the business objects are probably not expensive.

When testing the controller, we will almost certainly use a spy to represent the interactor because we don't want the calls to propagate to the database or presenter.

The presenter is interesting. We think of it as part of the GUI component, but in fact we're going to need a spy to test it. We don't want to test it with the real view, so we probably need a fifth component to hold the view apart from the controller and presenter.

That last little complication is common. We often modify our component boundaries because the tests demand it.

Conclusion（小结）

In this chapter, we looked at some of the more advanced aspects of TDD: from the incremental development of algorithms, to the problem of getting stuck, from the finite state machine nature of tests to test doubles and the TDD uncertainty principle. But we're not done. There's more. So, get a nice hot cup of tea and turn the improbability generator up to infinity.

4 TEST DESIGN

（设计）

If you look at the three laws of test-driven development (TDD), presented in Chapter 2, "Test-Driven Development," you could come to the conclusion that TDD is a shallow skill: Follow the three laws, and you are done. This is far from the truth. TDD is a deep skill. There are many layers to it; and they take months, if not years, to master.

In this section, we delve into just a few of those layers, ranging from various testing conundrums, such as databases and graphical user interfaces (GUIs), to the design principles that drive good test design, to patterns of testing, and to some interesting and profound theoretical possibilities.

TESTING DATABASES（测试数据库）

The first rule of testing databases is: *Don't test databases*. You don't need to test the database. You can assume that the database works. You'll find out soon enough if it doesn't.

What you really want to test are queries. Or, rather, you want to test that the commands that you send to the database are properly formed. If you write SQL directly, you are going to want to test that your SQL statements work as intended. If you use an object-relational mapping (ORM) framework, such as Hibernate, you are going to want to test that Hibernate operates the database the way you intended. If you use a NoSQL database, you are going to want to test that all your database accesses behave as you intended.

None of these tests require you to test business rules; they are only about the queries themselves. So, the second rule of testing databases is: *Decouple the database from the business rules*.

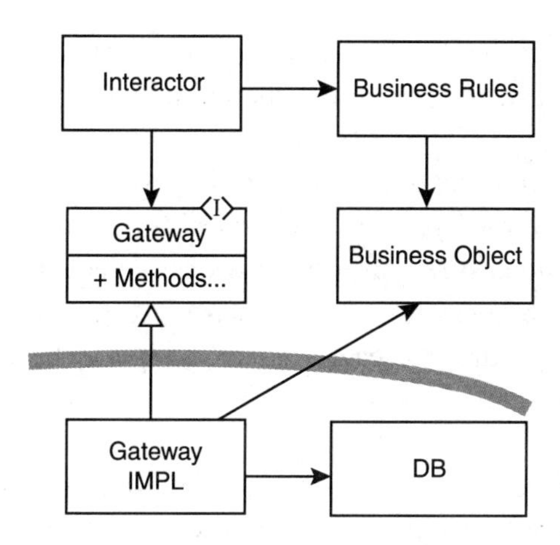

Figure 4.1 Testing the database

We decouple them by creating an interface, which I have called `Gateway`[1] in the diagram in Figure 4.1. Within the `Gateway` interface, we create one method for every kind of query we wish to perform. Those methods can take arguments that modify the query. For example, in order to fetch all the `Employees` from the database whose hiring date was after 2001, we might call the `Gateway` method `getEmployeesHiredAfter(2001)`.

Every query, update, delete, or add we want to perform in the database will have a corresponding method in a `Gateway` interface. There can, of course, be many `Gateways`, depending on how we want to partition the database.

The `GatewayImpl` class implements the gateway and directs the actual database to perform the functions required. If this is a SQL database, then all the SQL is created within the `GatewayImpl` class. If you are using ORM, the ORM framework is manipulated by the `GatewayImpl` class. Neither SQL nor the ORM framework nor the database API is known above the architectural boundary that separates the `Gateway` from the `GatewayImp`l.

Indeed, we don't even want the schema of the database known above that boundary. The `GatewayImpl` should unpack the rows, or data elements,

1. Martin Fowler, *Patterns of Enterprise Application Architecture* (Addison-Wesley, 2003), 466.

retrieved from the database and use that data to construct appropriate business objects to pass across the boundary to the business rules.

And now testing the database is trivial. You create a suitably simple test database, and then you call each query function of the `GatewayImpl` from your tests and ensure that it has the desired effect on that test database. Make sure that each query function returns a properly loaded set of business objects. Make sure that each update, add, and delete changes the database appropriately.

Do not use a production database for these tests. Create a test database with just enough rows to prove that the tests work and then make a backup of that database. Prior to running the tests, restore that backup so that the tests are *always* run against the same test data.

When testing the business rules, use stubs and spies to replace the `GatewayImpl` classes. Do not test business rules with the real database connected. This is slow and error prone. Instead, test that your business rules and interactors manipulate the `Gateway` interfaces correctly.

Testing GUIs（测试GUI）

The rules for testing GUIs are as follows:

1. Don't test GUIs.
2. Test everything but the GUI.
3. The GUI is smaller than you think it is.

Let's tackle the third rule first. The GUI is a lot smaller than you think it is. The GUI is just one very small element of the software that presents information on the screen. It is likely the smallest part of that software. It is the software that builds the commands that are sent to the engine that actually paints the pixels on the screen.

For a Web-based system, the GUI is the software that builds the HTML. For a desktop system, the GUI is the software that invokes the API of the graphic control software. Your job, as a software designer, is to make that GUI software as small as possible.

For example, does that software need to know how to format a date, or currency, or general numbers? No. Some other module can do that. All the GUI needs are the appropriate strings that represent the formatted dates, currencies, or numbers.

We call that other module a *presenter*. The presenter is responsible for formatting and arranging the data that is to appear on the screen, or in a window. The presenter does as much as possible toward that end, allowing us to make the GUI absurdly small.

So, for example, the presenter is the module that determines the state of every button and menu item. It specifies their names and whether or not they should be grayed out. If the name of a button changes based on the state of a window, it is the presenter that knows that state and changes that name. If a grid of numbers should appear on the screen, it is the presenter that creates a table of strings, all properly formatted and arranged. If there are fields that should have special colors or fonts, it is the presenter that determines those colors and fonts.

The presenter takes care of all of that detailed formatting and arrangement and produces a simple data structure full of strings and flags that the GUI can use to build the commands that get sent to the screen. And, of course, that makes the GUI very small indeed.

The data structure created by the presenter is often called a *view model*.

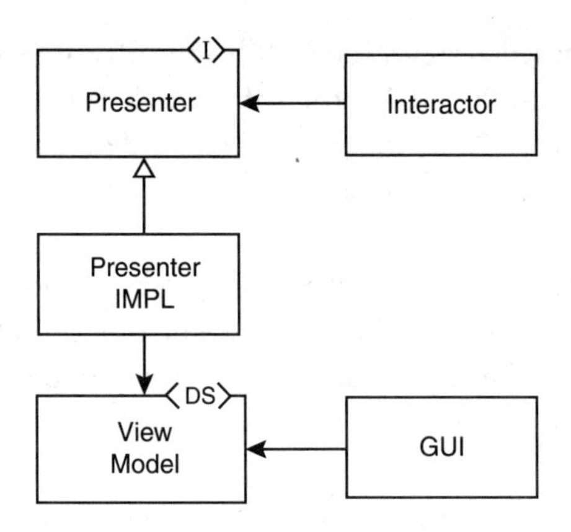

Figure 4.2 The interactor is responsible for telling the presenter what data should be presented to the screen.

In the diagram in Figure 4.2, the interactor is responsible for telling the presenter what data should be presented to the screen. This communication will be in the form of one or more data structures passed to the presenter through a set of functions. The actual presenter is shielded from the interactor by the presenter interface. This prevents the high-level interactor from depending on the implementation of the lower-level presenter.

The presenter builds the view model data structure, which the GUI then translates into the commands that control the screen.

Clearly, the interactor can be tested by using a spy for the presenter. Just as clearly, the presenter can be tested by sending it commands and inspecting the result in the view model.

The only thing that cannot be easily tested (with automated unit tests) is the GUI itself, and so we make it very small.

Of course, the GUI can still be tested; you just have to use your eyes to do it. But that turns out to be quite simple because you can simply pass a canned set of view models to the GUI and visually ensure that those view models are rendered appropriately.

There are tools that you can use to automate even that last step, but I generally advise against them. They tend to be slow and fragile. What's more, the GUI is most likely a very volatile module. Any time someone wants to change the look and appearance of something on the screen, it is bound to affect that GUI code. Thus, writing automated tests for that last little bit is often a waste of time because that part of the code changes so frequently, and for such evanescent reasons, any tests are seldom valid for long.

GUI Input（GUI 输入）

Testing GUI input follows the same rules: We drive the GUI to be as insignificant as possible. In the diagram in Figure 4.3, the GUI framework is the code that sits at the boundary of the system. It might be the Web container, or it might be something like Swing[2] or Processing[3] for controlling a desktop.

The GUI framework communicates with a controller through an `EventHandler` interface. This makes sure that the controller has no transitive source code dependency on the GUI framework. The job of the controller is to gather the necessary events from the GUI framework into a pure data structure that I have here called the `RequestModel`.

Once the `RequestModel` is complete, the controller passes it through the `InputBoundary` interface to the interactor. Again, the interface is there to ensure that the source code dependencies point in an architecturally sound direction.

2. https://docs.oracle.com/javase/8/docs/technotes/guides/swing/
3. https://processing.org/

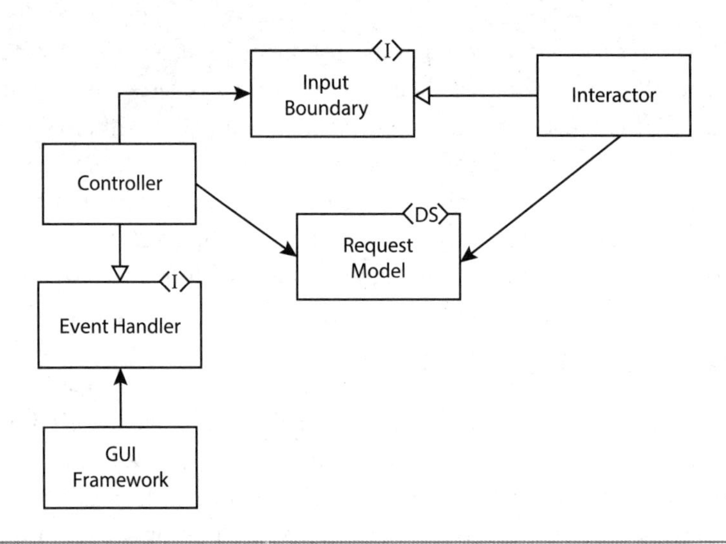

Figure 4.3 Testing the GUI

Testing the interactor is trivial; our tests simply create appropriate request models and hand them to the interactor. We can either check the results directly or use spies to check them. Testing the controller is also trivial—our tests simply invoke events through the event handler interface and then make sure that the controller builds the right request model.

Test Patterns（测试模式）

There are many different design patterns for testing, and there are several books that have documented them: *XUnit Test Patterns*[4] by Gerard Meszaros and *JUnit Recipes*[5] by J. B. Rainsberger and Scott Stirling, to mention just two of them.

It is not my intention to try to document all those patterns and recipes here. I just want to mention the three that I have found most useful over the years.

4. Gerard Meszaros, *XUnit Test Patterns: Refactoring Test Code* (Addison-Wesley, 2012).
5. J. B. Rainsberger and Scott Stirling, *JUnit Recipes: Practical Methods for Programmer Testing* (Manning, 2006).

TEST-SPECIFIC SUBCLASS（专为测试创建子类）

This pattern is primarily used as a safety mechanism. For example, let's say that you want to test the `align` method of the `XRay` class. However, the `align` method invokes the `turnOn` method. You probably don't want x-rays turned on every time you run the tests.

The solution, as shown in Figure 4.4, is to create a *test-specific subclass* of the `XRay` class that overrides the `turnOn` method to do nothing. The test creates an instance of the `SafeXRay` class and then calls the `assign` method, without having to worry that the x-ray machine will actually be turned on.

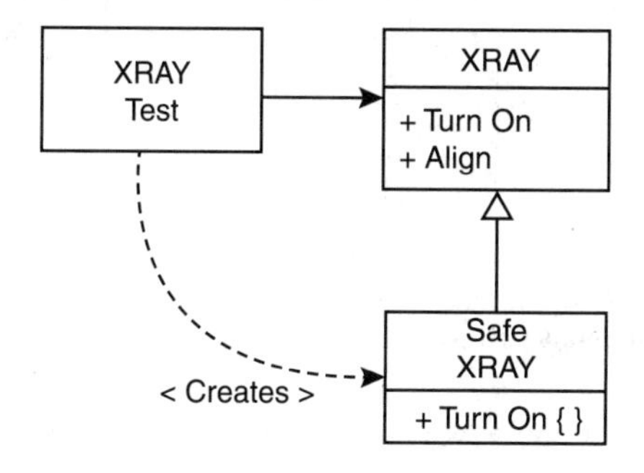

Figure 4.4 Test-Specific-Subclass pattern

It is often helpful to make the test-specific subclass a spy so that the test can interrogate the safe object about whether the unsafe method was actually called or not.

In the example, if `SafeXRay` were a spy, then the `turnOn` method would record its invocation, and the test method in the `XRayTest` class could interrogate that record to ensure that `turnOn` was actually called.

Sometimes the *Test-Specific Subclass* pattern is used for convenience and throughput rather than safety. For example, you may not wish the method being tested to start a new process or perform an expensive computation.

It is not at all uncommon for the dangerous, inconvenient, or slow operations to be extracted into new methods for the express purpose of overriding them in a test-specific subclass. This is just one of the ways that tests impact the design of the code.

SELF-SHUNT（自励）

A variation on that theme is the *Self-Shunt* pattern. Because the test class *is* a class, it is often convenient for the test class to become the test-specific subclass, as shown in Figure 4.5.

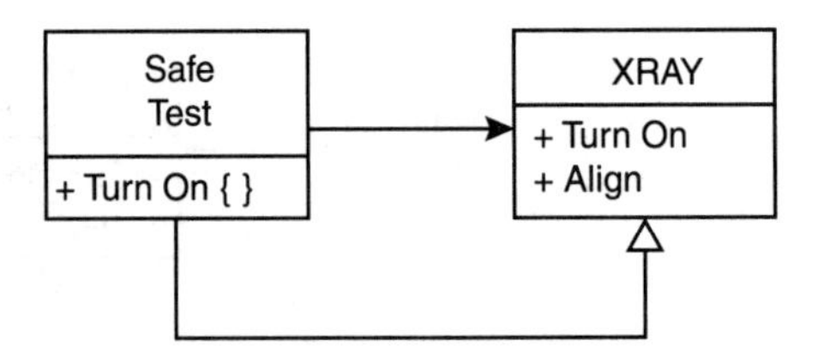

Figure 4.5 Self-Shunt pattern

In this case, it is the `XRayTest` class that overrides the `turnOn` method and can also act as the spy for that method.

I find Self-Shunt to be very convenient when I need a simple spy or a simple safety. On the other hand, the lack of a separate well-named class that specifically provides the safety or spying can be confusing for the reader, so I use this pattern judiciously.

It is important to remember, when using this pattern, that different testing frameworks construct the test classes at different times. For example, JUnit constructs a new instance of the test class for every test method invocation. NUnit, on the other hand, executes all test methods on a single instance of the test class. So, care must be taken to ensure that any spy variables are properly reset.

Humble Object

We like to think that every bit of code in the system can be tested using the three laws of TDD, but this is not entirely true. The parts of the code that communicate across a hardware boundary are perniciously difficult to test.

It is difficult, for example, to test what is displayed on the screen or what was sent out a network interface or what was sent out a parallel or serial I/O port. Without some specially designed hardware mechanisms that the tests can communicate with, such tests are impossible.

What's more, such hardware mechanisms may well be slow and/or unreliable. Imagine, for example, a video camera staring at the screen and your test code trying desperately to determine if the image coming back from the camera is the image you sent to the screen. Or imagine a loopback network cable that connects the output port of the network adapter to the input port. Your tests would have to read the stream of data coming in on that input port and look for the specific data you sent out on the output port.

In most cases, this kind of specialized hardware is inconvenient, if not entirely impractical.

The *Humble Object* pattern is a compromise solution. This pattern acknowledges that there is code that cannot be practicably tested. The goal of the pattern, therefore, is to *humiliate* that code by making it too simple to bother testing. We saw a simple example of this earlier, in the "Testing GUIs" section, but now let's take a deeper look.

The general strategy is shown in Figure 4.6. The code that communicates across the boundary is separated into two elements: the presenter and the *Humble Object* (denoted here as the `HumbleView`). The communication between the two is a data structure named the `Presentation`.

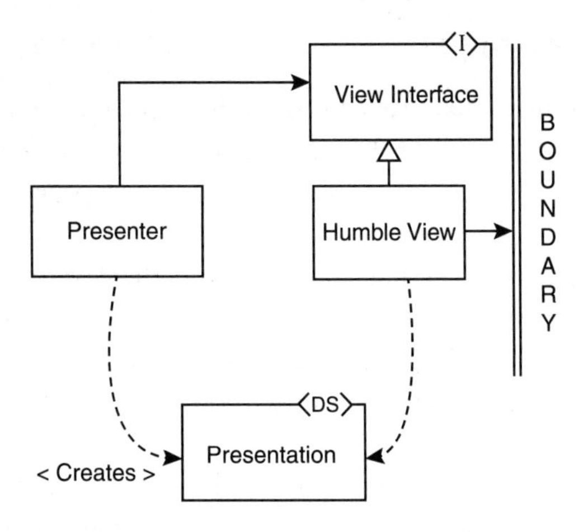

Figure 4.6 The general strategy

Let's assume that our application (not shown) wants to display something on the screen. It sends the appropriate data to the presenter. The presenter then unpacks that data into the simplest possible form and loads it into the `Presentation`. The goal of this unpacking is to eliminate all but the simplest processing steps from the `HumbleView`. The job of the `HumbleView` is simply to transport the unpacked data in the `Presentation` across the boundary.

To make this concrete, let's say that the application wants to put up a dialog box that has Post and Cancel buttons, a selection menu of order IDs, and a grid of dates and currency items. The data that the application sends to the presenter consists of that data grid, in the form of `Date` and `Money` objects. It also sends the list of selectable `Order` objects for the menu.

The presenter's job is to turn everything into strings and flags and load them into the `Presentation`. The `Money` and `Date` objects are converted into locale-specific strings. The `Order` objects are converted into ID strings. The names of the two buttons are loaded as strings. If one or more of the buttons should be grayed out, an appropriate flag is set in the `Presentation`.

The end result is that the `HumbleView` has nothing more to do than transport those strings across the boundary along with the metadata implied by the flags. Again, the goal is to make the `HumbleView` too simple to need testing.

It should be clear that this strategy will work for any kind of boundary crossing, not just displays.

To demonstrate, let's say we are coding the control software for a self-driving car. Let's also say that the steering wheel is controlled by a stepper motor that moves the wheel one degree per step. Our software controls the stepper motor by issuing the following command:

```
out(0x3ff9, d);
```

where `0x3ff9` is the IO address of the stepper motor controller, and `d` is `1` for right and `0` for left.

At the high level, our self-driving AI issues commands of this form to the `SteeringPresenter`:

```
turn(RIGHT, 30, 2300);
```

This means that the car (not the steering wheel!) should be gradually turned 30 degrees to the right over the next 2,300ms. To accomplish this, the wheel must be turned to the right a certain number of steps, at a certain rate, and then turned back to the left at a certain rate so that, after 2,300ms, the car is heading 30 degrees to the right of its previous course.

How can we test that the steering wheel is being properly controlled by the AI? We need to humiliate the low-level steering-wheel control software. We can do this by passing it a presentation, which is an array of data structures that looks like this:

```
struct SteeringPresentationElement{
  int steps;
  bool direction;
  int stepTime;
  int delay;
};
```

The low-level controller walks through the array and issues the appropriate number of `steps` to the stepper motor, in the specified `direction`, waiting `stepTime` milliseconds between each step and waiting `delay` milliseconds before moving to the next element in the array.

The `SteeringPresenter` has the task of translating the commands from the AI into the array of `SteeringPresentationElements`. In order to accomplish this, the `SteeringPresenter` needs to know the speed of the car and the ratio of the angle of the steering wheel to the angle of the wheels of the car.

It should be clear that the `SteeringPresenter` is easy to test. The test simply sends the appropriate `Turn` commands to the `SteeringPresenter` and then inspects the results in the resulting array of `SteeringPresentationElements`.

Finally, note the `ViewInterface` in the diagram. If we think of the `ViewInterface`, the presenter, and the `Presentation` as belonging together in a single component, then the `HumbleView` depends on that component. This is an architectural strategy for keeping the higher-level presenter from depending on the detailed implementation of the `HumbleView`.

TEST DESIGN（测试设计）

We are all familiar with the need to design our production code well. But have you ever thought about the design of your tests? Many programmers have not. Indeed, many programmers just throw tests at the code without any thought to how those tests should be designed. This always leads to problems.

THE FRAGILE TEST PROBLEM（脆弱测试问题）

One of the issues that plagues programmers who are new to TDD is the problem of fragile tests. A test suite is fragile when small changes to the production code cause many tests to break. The smaller the change to the production code, and the larger the number of broken tests, the more frustrating the issue becomes. Many programmers give up on TDD during their first few months because of this issue.

Fragility is always a design problem. If you make a small change to one module that forces many changes to other modules, you have an obvious design problem. In fact, breaking many things when something small changes *is the definition* of poor design.

Tests need to be designed just like any other part of the system. All the rules of design that apply to production code also apply to tests. Tests are not special in that regard. They must be properly designed in order to limit their fragility.

Much early guidance about TDD ignored the design of tests. Indeed, some of that guidance recommended structures that were counter to good design and led to tests that were tightly coupled to the production code and therefore very fragile.

The One-to-One Correspondence（一一对应）

One common and particularly detrimental practice is to create and maintain a one-to-one correspondence between production code modules and test modules. Newcomers to TDD are often erroneously taught that for every production module or class named χ, there should be a corresponding test module or class named `χTest`.

This, unfortunately, creates a powerful structural coupling between the production code and the test suite. That coupling leads to fragile tests. Any time the programmers want to change the module structure of the production code, they are forced to also change the module structure of the test code.

Perhaps the best way to see this structural coupling is visually (Figure 4.7).

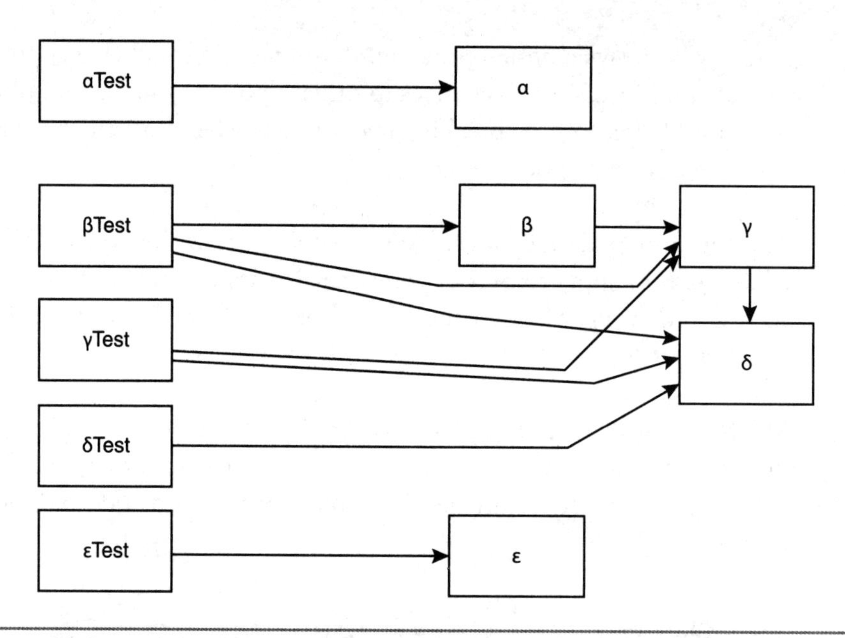

Figure 4.7 Structural coupling

On the right side of the diagram, we see five production code modules, α, β, γ, δ, and ε. The α and ε modules stand alone, but β is coupled to γ, which is coupled to δ. On the left, we see the test modules. Note that each of the test modules is coupled to the corresponding production code module. However, because β is coupled to γ and δ, `βTest` may also be coupled to γ and δ.

This coupling may not be obvious. The reason that `βTest` is likely to have couplings to γ and δ is that β may need to be constructed with γ and δ, or the methods of β may take γ and δ as arguments.

This powerful coupling between `βTest` and so much of the production code means that a minor change to δ could affect `βTest`, `γTest`, and `δTest`. Thus, the one-to-one correspondence between the tests and production code can lead to very tight coupling and fragility.

> *Rule 12: Decouple the structure of your tests from the structure of the production code.*

Breaking the Correspondence（打破对应关系）

To break, or avoid creating, the correspondence between tests and production code, we need to think of the test modules the way we think of all the other modules in a software system: as independent and decoupled from each other.

At first, this may seem absurd. You might argue that tests must be coupled to production code, because the tests *exercise* the production code. The last clause is true, but the predicate is false. Exercising code does not imply strong coupling. Indeed, good designers consistently strive to break strong couplings between modules while allowing those modules to interact with and exercise each other.

How is this accomplished? By creating interface layers.

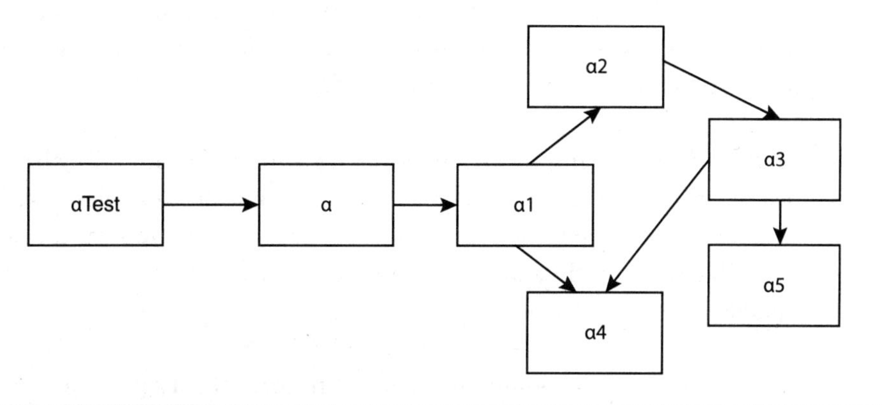

Figure 4.8 Interface layers

In the diagram in Figure 4.8, we see `αTest` coupled to α. Behind α we see a family of modules that support α but of which `αTest` is ignorant. The α module is the interface to that family. A good programmer is very careful to ensure that none of the details of the α family leak out of that interface.

As shown in the diagram in Figure 4.9, a disciplined programmer could protect `αTest` from the details within the α family by interposing a polymorphic interface between them. This breaks any transitive dependencies between the test module and the production code modules.

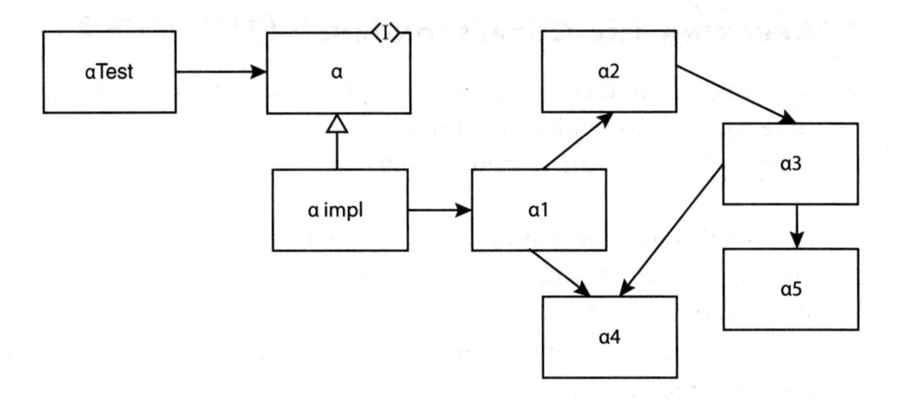

Figure 4.9 Interposing a polymorphic interface between the test and the α family

Again, this may seem absurd to the newcomer to TDD. How, you might ask, can we write tests against α5 when we cannot access that module from `αTest`? The answer to that question is simply that you do not need access to α5 in order to test the functionality of α5.

If α5 performs an important function for α, then that functionality must be testable through the α interface. That is not an arbitrary rule—it is a statement of mathematical certainty. If a behavior is important, it must also be visible through the interface. That visibility can be direct or indirect, but it must exist.

Perhaps an example would be beneficial to drive this point home.

The Video Store

The video store is a traditional example that demonstrates the concept of separating tests from production code quite well. Ironically, this example arose from an accident. The problem was first presented as a refactoring example in Martin Fowler's first edition of *Refactoring*.[6] Martin presented a rather ugly Java solution without tests and then proceeded to refactor the code into a much cleaner form.

6. Martin Fowler, *Refactoring* (Addison-Wesley, 1999).

In this example, we use TDD to create the program from scratch. You will learn the requirements by reading the tests as we go along.

Requirement 1: Regular movies rent, on the first day, for $1.50 and earn 1 renter point per day rented.

Red: We write a test class for the customer named `CustomerTest` and add the first test method.

```
public class CustomerTest {
  @Test
  public void RegularMovie_OneDay() throws Exception {
    Customer c = new Customer();
    c.addRental("RegularMovie", 1);
    assertEquals(1.5, c.getRentalFee(), 0.001);
    assertEquals(1, c.getRenterPoints());
  }
}
```

Green: We can make this pass trivially.

```
public class Customer {
  public void addRental(String title, int days) {
  }

  public double getRentalFee() {
    return 1.5;
  }

  public int getRenterPoints() {
    return 1;
  }
}
```

Refactor: We can clean this up quite a bit.

```
public class CustomerTest {
```

```
  private Customer customer;

  @Before
  public void setUp() throws Exception {
    customer = new Customer();
  }

  private void assertFeeAndPoints(double fee, int points) {
    assertEquals(fee, customer.getRentalFee(), 0.001);
    assertEquals(points, customer.getRenterPoints());
  }

  @Test
  public void RegularMovie_OneDay() throws Exception {
    customer.addRental("RegularMovie", 1);
    assertFeeAndPoints(1.5, 1);
  }
}
```

Requirement 2: The second and third days' rentals of regular movies are free, and no points are earned for them.

Green: No change to production code.

```
@Test
public void RegularMovie_SecondAndThirdDayFree() throws Exception {
  customer.addRental("RegularMovie", 2);
  assertFeeAndPoints(1.5, 1);
  customer.addRental("RegularMovie", 3);
  assertFeeAndPoints(1.5, 1);
}
```

Requirement 3: All subsequent days rent for $1.50 and earn 1 renter point.

Red: The test is simple.

```
@Test
public void RegularMovie_FourDays() throws Exception {
```

```
  customer.addRental("RegularMovie", 4);
  assertFeeAndPoints(3.0, 2);
}
```

Green: This isn't hard to fix.

```
public class Customer {
  private int days;

  public void addRental(String title, int days) {
    this.days = days;
  }

  public double getRentalFee() {
    double fee = 1.5;
    if (days > 3)
      fee += 1.5 * (days - 3);
    return fee;
  }

  public int getRenterPoints() {
    int points = 1;
    if (days > 3)
      points += (days - 3);
    return points;
  }
}
```

Refactor: There's a bit of duplication we can eliminate, but it causes some trouble.

```
public class Customer {
  private int days;

  public void addRental(String title, int days) {
    this.days = days;
  }
```

```
    public int getRentalFee() {
      return applyGracePeriod(150, 3);
    }

    public int getRenterPoints() {
      return applyGracePeriod(1, 3);
    }

    private int applyGracePeriod(int amount, int grace) {
      if (days > grace)
        return amount + amount * (days - grace);
      return amount;
    }
  }
```

Red: We want to use `applyGracePeriod` for both the renter points and the fee, but the fee is a `double`, and the points are an `int`. Money should never be a `double`! So, we changed the `fee` into an `int`, and all the tests broke. We need to go back and fix all our tests.

```
public class CustomerTest {
  private Customer customer;

  @Before
  public void setUp() throws Exception {
    customer = new Customer();
  }

  private void assertFeeAndPoints(int fee, int points) {
    assertEquals(fee, customer.getRentalFee());
    assertEquals(points, customer.getRenterPoints());
  }

  @Test
  public void RegularMovie_OneDay() throws Exception {
    customer.addRental("RegularMovie", 1);
    assertFeeAndPoints(150, 1);
  }
```

```
  @Test
  public void RegularMovie_SecondAndThirdDayFree() throws Exception {
    customer.addRental("RegularMovie", 2);
    assertFeeAndPoints(150, 1);
    customer.addRental("RegularMovie", 3);
    assertFeeAndPoints(150, 1);
  }

  @Test
  public void RegularMovie_FourDays() throws Exception {
    customer.addRental("RegularMovie", 4);
    assertFeeAndPoints(300, 2);
  }
}
```

Requirement 4: Children's movies rent for $1.00 per day and earn 1 point.

Red: The first day business rule is simple:

```
@Test
public void ChildrensMovie_OneDay() throws Exception {
  customer.addRental("ChildrensMovie", 1);
  assertFeeAndPoints(100, 1);
}
```

Green: It's not hard to make this pass with some very ugly code.

```
public int getRentalFee() {
  if (title.equals("RegularMovie"))
    return applyGracePeriod(150, 3);
  else
    return 100;
}
```

Refactor: But now we have to clean up that ugliness. There's no way the type of the video should be coupled to the title, so let's make a registry.

```
public class Customer {
  private String title;
  private int days;
  private Map<String, VideoType> movieRegistry = new HashMap<>();

  enum VideoType {REGULAR, CHILDRENS};

  public Customer() {
    movieRegistry.put("RegularMovie", REGULAR);
    movieRegistry.put("ChildrensMovie", CHILDRENS);
  }

  public void addRental(String title, int days) {
    this.title = title;
    this.days = days;
  }

  public int getRentalFee() {
    if (getType(title) == REGULAR)
      return applyGracePeriod(150, 3);
    else
      return 100;
  }

  private VideoType getType(String title) {
    return movieRegistry.get(title);
  }

  public int getRenterPoints() {
    return applyGracePeriod(1, 3);
  }

  private int applyGracePeriod(int amount, int grace) {
    if (days > grace)
      return amount + amount * (days - grace);
    return amount;
  }
}
```

That's better, but it violates the single responsibility principle[7] because the `Customer` class should not be responsible for initializing the registry. The registry should be initialized during early configuration of the system. Let's separate that registry from `Customer`:

```
public class VideoRegistry {
  public enum VideoType {REGULAR, CHILDRENS}

  private static Map<String, VideoType> videoRegistry =
                new HashMap<>();

  public static VideoType getType(String title) {
    return videoRegistry.get(title);
  }

  public static void addMovie(String title, VideoType type) {
    videoRegistry.put(title, type);
  }
}
```

`VideoRegistry` is a monostate[8] class, guaranteeing that there is only one instance. It is statically initialized by the test:

```
@BeforeClass
public static void loadRegistry() {
  VideoRegistry.addMovie("RegularMovie", REGULAR);
  VideoRegistry.addMovie("ChildrensMovie", CHILDRENS);
}
```

And this cleans up the `Customer` class a lot:

```
public class Customer {
  private String title;
```

7. Robert C. Martin, *Clean Architecture: A Craftsman's Guide to Software Structure and Design* (Addison-Wesley, 2018), 61ff.
8. Robert C. Martin, *Agile Software Development: Principles, Patterns, and Practices* (Prentice Hall, 2003), 180ff.

```
  private int days;

  public void addRental(String title, int days) {
    this.title = title;
    this.days = days;
  }

  public int getRentalFee() {
    if (VideoRegistry.getType(title) == REGULAR)
      return applyGracePeriod(150, 3);
    else
      return 100;
  }

  public int getRenterPoints() {
    return applyGracePeriod(1, 3);
  }

  private int applyGracePeriod(int amount, int grace) {
    if (days > grace)
      return amount + amount * (days - grace);
    return amount;
  }
}
```

Red: Note that requirement 4 said that customers earn 1 point for a children's movie, not 1 point per day. So, the next test looks like this:

```
@Test
public void ChildrensMovie_FourDays() throws Exception {
  customer.addRental("ChildrensMovie", 4);
  assertFeeAndPoints(400, 1);
}
```

I chose four days because of the `3` currently sitting as the second argument of the call to `applyGracePeriod` within the `getRenterPoints` method of the `Customer`. (Though we sometimes feign naïveté while doing TDD, we do actually know what's going on.)

Green: With the registry in place, this is easily repaired.

```
public int getRenterPoints() {
  if (VideoRegistry.getType(title) == REGULAR)
    return applyGracePeriod(1, 3);
  else
    return 1;
}
```

At this point, I want you to notice that there are no tests for the `VideoRegistry` class. Or, rather, no direct tests. `VideoRegistry` is, in fact, being tested indirectly because none of the passing tests would pass if `VideoRegistry` were not functioning properly.

Red: So far, our `Customer` class can handle only a single movie. Let's make sure it can handle more than one:

```
@Test
public void OneRegularOneChildrens_FourDays() throws Exception {
  customer.addRental("RegularMovie", 4); //$3+2p
  customer.addRental("ChildrensMovie", 4); //$4+1p

  assertFeeAndPoints(700, 3);
}
```

Green: That's just a nice little list and a couple of loops. It's also nice to move the registry stuff into the new `Rental` class:

```
public class Customer {
  private List<Rental> rentals = new ArrayList<>();

  public void addRental(String title, int days) {
    rentals.add(new Rental(title, days));
  }

  public int getRentalFee() {
    int fee = 0;
    for (Rental rental : rentals) {
```

```
      if (rental.type == REGULAR)
        fee += applyGracePeriod(150, rental.days, 3);
      else
        fee += rental.days * 100;
    }
    return fee;
  }

  public int getRenterPoints() {
    int points = 0;
    for (Rental rental : rentals) {
      if (rental.type == REGULAR)
        points += applyGracePeriod(1, rental.days, 3);
      else
        points++;
    }
    return points;
  }

  private int applyGracePeriod(int amount, int days, int grace) {
    if (days > grace)
      return amount + amount * (days - grace);
    return amount;
  }
}

public class Rental {
  public String title;
  public int days;
  public VideoType type;

  public Rental(String title, int days) {
    this.title = title;
    this.days = days;
    type = VideoRegistry.getType(title);
  }
}
```

This actually fails the old test because `Customer` now sums up the two rentals:

```
  @Test
  public void RegularMovie_SecondAndThirdDayFree() throws Exception {
    customer.addRental("RegularMovie", 2);
    assertFeeAndPoints(150, 1);
    customer.addRental("RegularMovie", 3);
    assertFeeAndPoints(150, 1);
  }
```

We have to divide that test in two. That's probably better anyway.

```
@Test
public void RegularMovie_SecondDayFree() throws Exception {
  customer.addRental("RegularMovie", 2);
  assertFeeAndPoints(150, 1);
}

@Test
public void RegularMovie_ThirdDayFree() throws Exception {
  customer.addRental("RegularMovie", 3);
  assertFeeAndPoints(150, 1);
}
```

Refactor: There's an awful lot I don't like about the `Customer` class now. Those two ugly loops with the strange `if` statements inside them are pretty awful. We can extract a few nicer methods from those loops.

```
public int getRentalFee() {
  int fee = 0;
  for (Rental rental : rentals)
    fee += feeFor(rental);
  return fee;
}

private int feeFor(Rental rental) {
  int fee = 0;
  if (rental.getType() == REGULAR)
```

```
      fee += applyGracePeriod(150, rental.getDays(), 3);
    else
      fee += rental.getDays() * 100;
    return fee;
  }

  public int getRenterPoints() {
    int points = 0;
    for (Rental rental : rentals)
      points += pointsFor(rental);
    return points;
  }

  private int pointsFor(Rental rental) {
    int points = 0;
    if (rental.getType() == REGULAR)
      points += applyGracePeriod(1, rental.getDays(), 3);
    else
      points++;
    return points;
  }
```

Those two private functions seem to play more with the `Rental` than with the `Customer`. Let's move them along with their utility function `applyGracePeriod`. This makes the `Customer` class much cleaner.

```
public class Customer {
  private List<Rental> rentals = new ArrayList<>();

  public void addRental(String title, int days) {
    rentals.add(new Rental(title, days));
  }

  public int getRentalFee() {
    int fee = 0;
    for (Rental rental : rentals)
      fee += rental.getFee();
    return fee;
  }
```

```
  public int getRenterPoints() {
    int points = 0;
    for (Rental rental : rentals)
      points += rental.getPoints();
    return points;
  }
}
```

The Rental class has grown a lot and is much uglier now:

```
public class Rental {
  private String title;
  private int days;
  private VideoType type;

  public Rental(String title, int days) {
    this.title = title;
    this.days = days;
    type = VideoRegistry.getType(title);
  }

  public String getTitle() {
    return title;
  }

  public VideoType getType() {
    return type;
  }

  public int getFee() {
    int fee = 0;
    if (getType() == REGULAR)
      fee += applyGracePeriod(150, days, 3);
    else
      fee += getDays() * 100;
    return fee;
  }
```

```
  public int getPoints() {
    int points = 0;
    if (getType() == REGULAR)
      points += applyGracePeriod(1, days, 3);
    else
      points++;
    return points;
  }

  private static int applyGracePeriod(int amount, int days, int grace)
  {
    if (days > grace)
      return amount + amount * (days - grace);
    return amount;
  }
}
```

Those ugly `if` statements need to be gotten rid of. Every new type of video is going to mean another clause in those statements. Let's head that off with some subclasses and polymorphism.

First, there's the abstract `Movie` class. It's got the `applyGracePeriod` utility and two abstract functions to get the fee and the points.

```
public abstract class Movie {
  private String title;

  public Movie(String title) {
    this.title = title;
  }

  public String getTitle() {
    return title;
  }

  public abstract int getFee(int days, Rental rental);
  public abstract int getPoints(int days, Rental rental);
```

```
  protected static int applyGracePeriod(int amount, int days,
                                        int grace) {
    if (days > grace)
      return amount + amount * (days - grace);
    return amount;
  }
}
```

`RegularMovie` is pretty simple:

```
public class RegularMovie extends Movie {
  public RegularMovie(String title) {
    super(title);
  }

  public int getFee(int days, Rental rental) {
    return applyGracePeriod(150, days, 3);
  }

  public int getPoints(int days, Rental rental) {
    return applyGracePeriod(1, days, 3);
  }
}
```

`ChildrensMovie` is even simpler:

```
public class ChildrensMovie extends Movie {
  public ChildrensMovie(String title) {
    super(title);
  }

  public int getFee(int days, Rental rental) {
    return days * 100;
  }

  public int getPoints(int days, Rental rental) {
    return 1;
  }
}
```

There's not much left of `Rental`—just a couple of delegator functions:

```
public class Rental {
  private int days;
  private Movie movie;

  public Rental(String title, int days) {
    this.days = days;
    movie = VideoRegistry.getMovie(title);
  }

  public String getTitle() {
    return movie.getTitle();
  }

  public int getFee() {
    return movie.getFee(days, this);
  }

  public int getPoints() {
    return movie.getPoints(days, this);
  }
}
```

The `VideoRegistry` class turned into a factory for `Movie`.

```
public class VideoRegistry {
  public enum VideoType {REGULAR, CHILDRENS;}

  private static Map<String, VideoType> videoRegistry =
                 new HashMap<>();

  public static Movie getMovie(String title) {
    switch (videoRegistry.get(title)) {
      case REGULAR:
```

```
        return new RegularMovie(title);
      case CHILDRENS:
        return new ChildrensMovie(title);
    }
    return null;
  }

  public static void addMovie(String title, VideoType type) {
    videoRegistry.put(title, type);
  }
}
```

And Customer? Well, it just had the wrong name all this time. It is really the RentalCalculator class. It is the class that protects our tests from the family of classes that serves it.

```
public class RentalCalculator {
  private List<Rental> rentals = new ArrayList<>();

  public void addRental(String title, int days) {
    rentals.add(new Rental(title, days));
  }

  public int getRentalFee() {
    int fee = 0;
    for (Rental rental : rentals)
      fee += rental.getFee();
    return fee;
  }

  public int getRenterPoints() {
    int points = 0;
    for (Rental rental : rentals)
      points += rental.getPoints();
    return points;
  }
}
```

Now let's look at a diagram of the result (Figure 4.10).

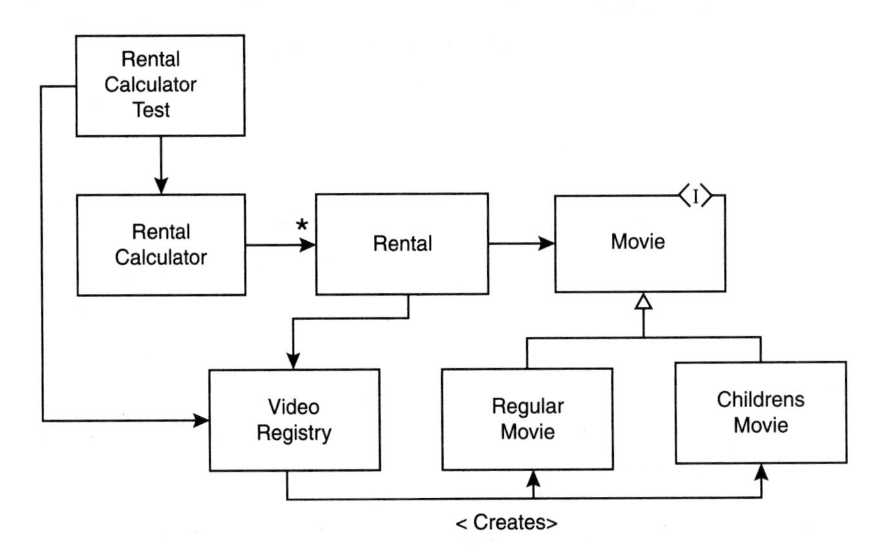

Figure 4.10 The result

As the code evolved, all those classes to the right of `RentalCalculator` were created by various refactorings. Yet `RentalCalculatorTest` knows nothing of them other than `VideoRegistry`, which it must initialize with test data. Moreover, no other test module exercises those classes. `RentalCalculatorTest` tests all those other classes indirectly. The one-to-one correspondence is broken.

This is the way that good programmers protect and decouple the structure of the production code from the structure of the tests, thereby avoiding the fragile test problem.

In large systems, of course, this pattern will repeat over and over. There will be many families of modules, each protected from the test modules that exercise them by their own particular facades or interface modules.

Some might suggest that tests that operate a family of modules through a facade are integration tests. We talk about integration tests later in this book. For now, I'll simply point out that the purpose of integration tests is very different from the purpose of the tests shown here. These are *programmer tests,* tests written by programmers for programmers for the purpose of specifying behavior.

Specificity versus Generality（具体 VS 通用）

Tests and production code must be decoupled by yet another factor that we learned about in Chapter 2, when we studied the prime factors example. I wrote it as a mantra in that chapter. Now I'll write it as a rule.

> *Rule 13: As the tests get more specific, the code gets more generic.*

The family of production code modules grows as the tests grow. However, they evolve in very different directions.

As each new test case is added, the suite of tests becomes increasingly specific. However, programmers should drive the family of modules being tested in the opposite direction. That family should become increasingly general (Figure 4.11).

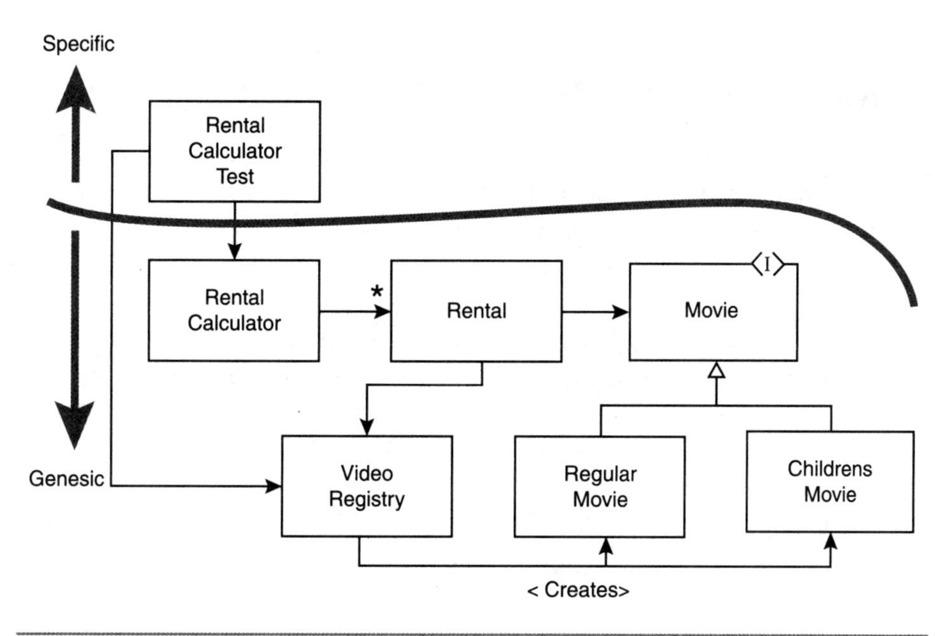

Figure 4.11 The suite of tests becomes more specific, while the family of modules being tested becomes more general.

This is one of the goals of the refactoring step. You saw it happening in the video store example. First a test case was added. Then some ugly production code was added to get the test to pass. That production code was not general. Often, in fact, it was deeply specific. Then, in the refactoring step, that specific code was massaged into a more general form.

This divergent evolution of the tests and the production code means that the shapes of the two will be remarkably different. The tests will grow into a linear list of constraints and specifications. The production code, on the other hand, will grow into a rich family of logic and behavior organized to address the underlying abstraction that drives the application.

This divergent style further decouples the tests from the production code, protecting the two from changes in the other.

Of course, the coupling can never be completely broken. There will be changes in one that force changes in the other. The goal is not to eliminate such changes but to minimize them. And the techniques described are effective toward that end.

Transformation Priority Premise（转换优先顺序）

The previous chapters have led up to a fascinating observation. When we practice the discipline of TDD, we incrementally make the tests more specific, while we manipulate the production code to be ever more general. But how do these changes take place?

Adding a constraint to a test is a simple matter of either adding a new assertion to an existing test or adding a whole new test method to arrange, act, and then assert the new constraint. This operation is entirely additive. No existing test code is changed. New code is added.

Making the new constraint pass the tests, however, is very often not an additive process. Instead, the existing production code must be transformed to behave differently. These transformations are small changes to the existing code that alter the behavior of that code.

Then, of course, the production code is refactored in order to clean it up. These refactorings are also small changes to the production code, but in this case, they preserve the behavior.

Already you should see the correlation to the Red/Green/Refactor loop. The Red step is additive. The Green step is transformative. The Blue step is restorative.

We discuss restorative refactorings in Chapter 5, “Refactoring.” Here, we discuss the transformations.

Transformations are small changes to the code that change behavior and simultaneously generalize the solution. The best way to explain this is with an example.

Recall the prime factors kata from Chapter 2. Early on, we saw a failing test and a degenerate implementation.

```
public class PrimeFactorsTest {
  @Test
  public void factors() throws Exception {
    assertThat(factorsOf(1), is(empty()));
  }

  private List<Integer> factorsOf(int n) {
    return null;
  }
}
```

We made the failing test pass by transforming the `null` into `new ArrayList<>()`, as follows:

```
private List<Integer> factorsOf(int n) {
  return new ArrayList<>();
}
```

That transformation changed the behavior of the solution but also generalized it. That `null` was extremely specific. `ArrayList` is more general than `null`.

The next failing test case also resulted in generalizing transformations:

```
assertThat(factorsOf(2), contains(2));

private List<Integer> factorsOf(int n) {
  ArrayList<Integer> factors = new ArrayList<>();
  if (n>1)
    factors.add(2);
  return factors;
}
```

First, `ArrayList` was extracted to the `factors` variable, and then the `if` statement was added. Both of these transformations are generalizing. Variables are always more general than constants. However, the `if` statement is only partially generalizing. It is specific to the test because of the `1` and the `2`, but it softens that specificity with the `n>1` inequality. That inequality remained part of the general solution all the way to the end.

Armed with this knowledge, let's look at some other transformations.

{} → Nil（无代码→空值）

This is usually the very first transformation employed at the start of a TDD session. We begin with no code at all. We write the most degenerate test we can think of. Then, to get it to compile and fail, we make the function we are testing return `null`,[9] as we did in the prime factors kata.

```
private List<Integer> factorsOf(int n) {
  return null;
}
```

9. Or the most degenerate allowed return value.

This code transforms nothing into a function that returns nothing. Doing so seldom makes the failing test pass, so the next transformation usually follows immediately.

Nil → Constant（空值→常量）

Again, we see this in the prime factors kata. The `null` we returned is transformed into an empty list of integers.

```
private List<Integer> factorsOf(int n) {
  return new ArrayList<>();
}
```

We also saw this in the bowling game kata in Chapter 2, though in that case, we skipped the {} → Nil transformation and went straight to the constant.

```
  public int score() {
    return 0;
  }
```

Constant → Variable

This transformation changes a constant into a variable. We saw this in the stack kata (Chapter 2) when we created the `empty` variable to hold the `true` value that `isEmpty` had been returning.

```
public class Stack {
  private boolean empty = true;

  public boolean isEmpty() {
    return empty;
  }
      . . .
}
```

We saw this again in prime factors when, in order to pass the case for factoring `3`, we replaced the constant `2` with the argument `n`.

```
private List<Integer> factorsOf(int n) {
  ArrayList<Integer> factors = new ArrayList<>();
  if (n>1)
    factors.add(n);
  return factors;
}
```

At this point, it should be obvious that every one of these transformations, so far, moves the code from a very specific state to a slightly more general state. Each of them is a generalization, a way to make the code handle a wider set of constraints than before.

If you think about it carefully, you'll realize that each of these transformations widens the possibilities much more than the constraint placed on the code by the currently failing test. Thus, as these transformations are applied, one by one, the race between the constraints of the tests and the generality of the code must end in favor of the generalizations. Eventually, the production code will become so general that it will pass all future constraints within the current requirements.

But I digress.

Unconditional → Selection（无条件→条件选择）

This transformation adds an `if` statement, or the equivalent. This is not always a generalization. Programmers must take care not to make the predicate of the selection specific to the currently failing test.

We saw this transformation in the prime factors kata when we needed to factor the number 2. Note that the predicate of the `if` statement in that kata was not `(n==2)`; that would have been too specific. The `(n>1)` inequality was an attempt to keep the `if` statement more general.

```
private List<Integer> factorsOf(int n) {
  ArrayList<Integer> factors = new ArrayList<>();
  if (n>1)
    factors.add(2);
```

```
  return factors;
}
```

VALUE → LIST（值→列表）

This generalizing transformation changes a variable that holds a single value into a list of values. The list could be an array or a more complex container. We saw this transformation in the stack kata when we changed the `element` variable into the `elements` array.

```
public class Stack {
  private int size = 0;
  private int[] elements = new int[2];

  public void push(int element) {
    this.elements[size++] = element;
  }

  public int pop() {
    if (size == 0)
      throw new Underflow();
    return elements[--size];
  }
}
```

STATEMENT → RECURSION（语句→递归）

This generalizing transformation changes a single statement into a recursive statement, in lieu of a loop. These kinds of transformations are very common in languages that support recursion, especially those like Lisp and Logo that have no looping facilities *other* than recursion. The transformation changes an expression that is evaluated once into an expression that is evaluated in terms of itself. We saw this transformation in the word-wrap kata in Chapter 3, "Advanced TDD."

```
private String wrap(String s, int w) {
  if (w >= s.length())
    return s;
```

```
  else
    return s.substring(0, w) + "\n" + wrap(s.substring(w), w);
}
```

Selection → Iteration（条件选择→遍历）

We saw this several times in the prime factors kata when we converted those `if` statements into `while` statements. This is clearly a generalization because iteration is the general form of selection, and selection is merely degenerate iteration.

```
private List<Integer> factorsOf(int n) {
  ArrayList<Integer> factors = new ArrayList<>();
  if (n > 1) {
    while (n % 2 == 0) {
      factors.add(2);
      n /= 2;
    }
  }
  if (n > 1)
    factors.add(n);
  return factors;
}
```

Value → Mutated Value（值→改变了的值）

This transformation mutates the value of a variable, usually for the purpose of accumulating partial values in a loop or incremental computation. We saw this in in several of the katas but perhaps most significantly in the sort kata in Chapter 3.

Note that the first two assignments are not mutations. The `first` and `second` values are simply initialized. It is the `list.set(...)` operations that are the mutations. They actually change the elements within the list.

```
private List<Integer> sort(List<Integer> list) {
  if (list.size() > 1) {
```

```
    if (list.get(0) > list.get(1)) {
      int first = list.get(0);
      int second = list.get(1);
      list.set(0, second);
      list.set(1, first);
    }
  }
  return list;
}
```

Example: Fibonacci（示例：斐波那契数列）

Let's try a simple kata and keep track of the transformations. We'll do the tried and true Fibonacci kata. Remember that `fib(0)` = 1, `fib(1)` = 1, and `fib(n)` = `fib(n-1)` + `fib(n-2)`.

We begin, as always, with a failing test. If you are wondering why I'm using `BigInteger`, it is because Fibonacci numbers get big very quickly.

```
public class FibTest {
  @Test
  public void testFibs() throws Exception {
    assertThat(fib(0), equalTo(BigInteger.ONE));
  }

  private BigInteger fib(int n) {
    return null;
  }
}
```

We can make this pass by using the Nil → Constant transformation.

```
private BigInteger fib(int n) {
  return new BigInteger("1");
}
```

Yes, I thought the use of the `String` argument was odd too; but that's the Java library for you.

The next test passes right out of the box:

```
@Test
public void testFibs() throws Exception {
  assertThat(fib(0), equalTo(BigInteger.ONE));
  assertThat(fib(1), equalTo(BigInteger.ONE));
}
```

The next test fails:

```
@Test
public void testFibs() throws Exception {
  assertThat(fib(0), equalTo(BigInteger.ONE));
  assertThat(fib(1), equalTo(BigInteger.ONE));
  assertThat(fib(2), equalTo(new BigInteger("2")));
}
```

We can make this pass by using Unconditional → Selection:

```
private BigInteger fib(int n) {
  if (n > 1)
    return new BigInteger("2");
  else
    return new BigInteger("1");
}
```

This is perilously close to being more specific than general, though it titillates me with the potential for negative arguments to the `fib` function.

The next test tempts us to go for the gold:

```
assertThat(fib(3), equalTo(new BigInteger("3")));
```

The solution uses Statement → Recursion:

```
private BigInteger fib(int n) {
  if (n > 1)
```

```
    return fib(n-1).add(fib(n-2));
  else
    return new BigInteger("1");
}
```

This is a very elegant solution. It's also horrifically expensive in terms of time[10] and memory. Going for the gold too early often comes at a cost. Is there another way we could have done that last step?

Of course there is:

```
private BigInteger fib(int n) {
  return fib(BigInteger.ONE, BigInteger.ONE, n);
}

private BigInteger fib(BigInteger fm2, BigInteger fm1, int n) {
  if (n>1)
    return fib(fm1, fm1.add(fm2), n-1);
  else
    return fm1;

}
```

This is a nice tail-recursive algorithm that is tolerably fast.[11]

You might think that last transformation was just a different application of Statement → Recursion, but it wasn't. It was actually Selection → Iteration. In fact, if the Java compiler would deign to offer us tail-call-optimization,[12] it would translate almost exactly to the following code. Note the implied `if->while`.

```
private BigInteger fib(int n) {
  BigInteger fm2 = BigInteger.ONE;
  BigInteger fm1 = BigInteger.ONE;
```

---

10. `fib(40)==165580141` took nine seconds to compute on my 2.3GHz MacBook Pro.
11. `fib(100)==573147844013817084101` in 10ms.
12. Java, Java, wherefore art thou Java?

```
    while (n>1) {
      BigInteger f = fm1.add(fm2);
      fm2 = fm1;
      fm1 = f;
      n--;
    }
    return fm1;
  }
```

I took you on that little diversion to make an important point:

> *Rule 14: If one transformation leads you to a suboptimal solution, try a different transformation.*

This is actually the second time we have encountered a situation in which a transformation led us to a suboptimal solution and a different transformation produced much better results. The first time was back in the sort kata. In that case, it was the Value → Mutated Value transformation that led us astray and drove us to implement a bubble sort. When we replaced that transformation with Unconditional → Selection, we wound up implementing a quick sort. Here was the critical step:

```
private List<Integer> sort(List<Integer> list) {
  if (list.size() <= 1)
    return list;
  else {
    int first = list.get(0);
    int second = list.get(1);
    if (first > second)
      return asList(second, first);
    else
      return asList(first, second);
  }
}
```

The Transformation Priority Premise（变换模式优先顺序假设）

As we have seen, there are sometimes forks in the road as we are following the three laws of TDD. Each tine of a fork uses a different transformation to make the currently failing test pass. When faced with such a fork, is there a way to choose the best transformation to use? Or, to say this differently, is one transformation better than another in every case? Is there a *priority* to the transformations?

I believe there is. I'll describe that priority to you in just a moment. However, I want to make it clear that this belief of mine is only a *premise*. I have no mathematical proof, and I am not sure that it holds in every case. What I am relatively certain of is that you are likely to wind up at better implementations if you choose the transformations in something like the following order:

- {} → Nil
- Nil → Constant
- Constant → Variable
- Unconditional → Selection
- Value → List
- Selection → Iteration
- Statement → Recursion
- Value → Mutated Value

Don't make the mistake of thinking that the order here is natural and immune to violation (e.g., that Constant → Variable cannot be used until Nil → Constant has been completed). Many programmers might make a test pass by transforming Nil to a *Selection* of two constants without going through the Nil → Constant step.

In other words, if you are tempted to pass a test by combining two or more transformations, you may be missing one or more tests. Try to find a test that can be passed by using just one of these transformations. Then, when you find yourself at a fork in the road, first choose the tine of that fork that can be passed by using the transformation that is *higher* on the list.

Does this mechanism always work? Probably not, but I've had pretty good luck with it. And as we've seen, it gave us better results for both the sort and the Fibonacci algorithms.

Astute readers will have realized by now that following the transformations in the specified order will lead you to implement solutions using the functional programming style.

Conclusion（小结）

This concludes our discussion of the discipline of TDD. We've covered a lot of ground in the last three chapters. In this chapter, we talked about the problems and patterns of test design. From GUIs to databases, from specifications to generalities, and from transformations to priorities.

But, of course, we're not done. There's the fourth law to consider: refactoring. That's the topic of the next chapter.

5 REFACTORING

（重构）

In 1999, I read *Refactoring*[1] by Martin Fowler. It is a classic, and I encourage you to get a copy and read it. He has recently published a second edition,[2] which has been considerably rewritten and modernized. The first edition presents examples in Java; the second edition presents examples in JavaScript.

At the time that I was reading the first edition, my twelve-year-old son, Justin, was on a hockey team. For those of you who are not hockey parents, the games involve five minutes of your child playing on the ice and ten to fifteen minutes off the ice so that they can cool down.

While my son was off the ice, I read Martin's wonderful book. It was the first book I had ever read that presented code as something *malleable*. Most other books of the period, and before, presented code in final form. But *this* book showed how you could take bad code and clean it.

As I read it, I would hear the crowd cheer for the kids on the ice, and I would cheer along with them—but I was not cheering for the game. I was cheering for what I was reading in that book. In many ways, it was the book that put me on the path to writing *Clean Code*.[3]

Nobody said it better than Martin:

> *Any fool can write code that a computer can understand. Good programmers write code that humans can understand.*

This chapter presents the art of refactoring from my personal point of view. It is not intended as a replacement for Martin's book.

---

1. Martin Fowler, *Refactoring: Improving the Design of Existing Code*, 1st ed. (Addison-Wesley, 1999).
2. Martin Fowler, *Refactoring: Improving the Design of Existing Code,* 2nd ed. (Addison-Wesley, 2019).
3. Robert C. Martin, *Clean Code* (Addison-Wesley, 2009).

What Is Refactoring?（什么是重构）

This time, I paraphrase Martin with my own quote:

> *Refactoring is a sequence of small changes that improve the structure of the software without changing its behavior—as proven by passing a comprehensive suite of tests after each change in the sequence.*

There are two critical points in this definition.

First, refactoring *preserves* behavior. After a refactoring, or a sequence of refactorings, the behavior of the software remains unchanged. The only way I know to prove the preservation of behavior is to consistently pass a suite of *comprehensive* tests.

Second, each individual refactoring is *small*. How small? I have a rubric: *small enough that I won't have to debug.*

There are many specific refactorings, and I describe some of them in the pages that follow. There are many other changes to code that are not part of the refactoring canon but that are still behavior-preserving changes to structure. Some refactorings are so formulaic that your IDE can do them for you. Some are simple enough that you can do them manually without fear. Some are a bit more involved and require significant care. For those in the latter case, I apply my rubric. If I fear that I will wind up in a debugger, I break the change down into smaller, safer pieces. If I wind up in a debugger anyway, I adjust my fear threshold in favor of caution.

> *Rule 15: Avoid using debuggers.*

The purpose of refactoring is to clean the code. The process for refactoring is the red → green → refactor cycle. Refactoring is a constant activity, not a scheduled and planned activity. You keep the code clean by refactoring it every time around the red → green → refactor loop.

There will be times when larger refactorings are necessary. You will inevitably find that the design for your system needs updating, and you'll want to make that design change throughout the body of the code. You do not schedule this. You do not stop adding features and fixing bugs to do it. You simply add a bit of extra refactoring effort to the red → green → refactor cycle and gradually make the desired changes as you also continuously deliver business value.

The Basic Toolkit（基础工具包）

I use a few refactorings much more than any of the others. They are automated by the IDE that I use. I urge you to learn these refactorings by heart and understand the intricacies of your IDE's automation of them.

Rename（重命名）

A chapter in my *Clean Code* book discusses how to name things well. There are many other references[4] for learning to name things well. The important thing is . . . to name things well.

Naming things is hard. Finding the right name for something is often a process of successive, iterative improvements. Do not be afraid to pursue the right name. Improve names as often as you can when the project is young.

As the project ages, changing names becomes increasingly difficult. Increasing numbers of programmers will have committed the names to memory and will not react well if those names are changed without warning. As time goes on, renaming important classes and functions will require meetings and consensus.

So, as you write new code, and while that code is not too widely known, experiment with names. Rename your classes and methods frequently. As you do, you'll find that you'll want to group them differently. You'll move methods from one class to another to remain consistent with your new names. You'll change the partitioning of functions and classes to correspond to the new naming scheme.

---

4. Another good reference is *Domain-Driven Design: Tackling Complexity in the Heart of Software* by Eric Evans (Addison-Wesley, 2013).

In short, the practice of searching for the best names will likely have a profoundly positive effect on the way you partition the code into classes and modules.

So, learn to use the **Rename** refactoring frequently and well.

EXTRACT METHOD（方法抽取）

The **Extract Method** refactoring may be the most important of all the refactorings. Indeed, this refactoring may be the most important mechanism for keeping your code clean and well organized.

My advice is to follow the *extract 'til you drop* discipline.

This discipline pursues two goals. First, every function should do *one thing*.[5] Second, your code should read like *well-written prose*.[6]

A function does *one thing* when no other function can be extracted from it. Therefore, in order that your functions all do *one thing*, you should extract and extract and extract until you cannot extract any more.

This will, of course, lead to a plethora of little tiny functions. And this may disturb you. You may feel that so many little tiny functions will obscure the intent of your code. You may worry that it would be easy to get lost within such a huge swarm of functions.

But the opposite happens. The intent of your code becomes much more obvious. The levels of abstraction become crisp and the lines between them clear.

Remember that languages nowadays are rich with modules, classes, and namespaces. This allows you to build a hierarchy of names within which to place your functions. Namespaces hold classes. Classes hold functions. Public functions refer to private functions. Classes hold inner and nested classes.

---

5. Martin, *Clean Code*, p. 7.
6. Martin, p. 8.

And so on. Take advantage of these tools to create a structure that makes it easy for other programmers to locate the functions you have written.

And then choose good names. Remember that the length of a function's name should be *inversely* proportional to the scope that contains it. The names of public functions should be relatively short. The names of private functions should be longer.

As you extract and extract, the names of the functions will get longer and longer because the purpose of the function will become less and less general. Most of these extracted functions will be called from only one place, so their purpose will be extremely specialized and precise. The names of such specialized and precise functions must be long. They will likely be full clauses or even sentences.

These functions will be called from within the parentheses of `while` loops and `if` statements. They will be called from within the bodies of those statements as well, leading to code that looks like this:

```
if (employeeShouldHaveFullBenefits())
  AddFullBenefitsToEmployee();
```

It will make your code read like *well-written prose*.

Using the **Extract Method** refactoring is also how you will get your functions to follow *the stepdown rule*.[7] We want each line of a function to be at the same level of abstraction, and that level should be one level below the *name* of the function. To achieve this, we extract all code snippets within a function that are below the desired level.

Extract Variable（变量抽取）

If **Extract Method** is the most important of refactorings, **Extract Variable** is its ready assistant. It turns out that in order to extract methods, you often must extract variables first.

7. Martin, p. 37.

For example, consider this refactoring from the bowling game in Chapter 2, "Test-Driven Development." We started with this:

```
@Test
public void allOnes() throws Exception {
  for (int i=0; i<20; i++)
    g.roll(1);
  assertEquals(20, g.score());
}
```

And we ended up with this:

```
private void rollMany(int n, int pins) {
  for (int i = 0; i < n; i++) {
    g.roll(pins);
  }
}

@Test
public void allOnes() throws Exception {
  rollMany(20, 1);
  assertEquals(20, g.score());
}
```

The sequence of refactorings was as follows:

1. **Extract Variable:** The `1` in `g.roll(1)` was extracted into a variable named `pins`.
2. **Extract Variable:** The `20` in `assertEquals(20, g.score());` was extracted into a variable named `n`.
3. The two variables were moved above the `for` loop.
4. **Extract Method:** The `for` loop was extracted into the `rollMany` function. The names of the variables became the names of the arguments.
5. **Inline:** The two variables were inlined. They had served their purpose and were no longer needed.

Another common usage for **Extract Variable** is to create an *explanatory variable*.[8] For example, consider the following `if` statement:

```
if (employee.age > 60 && employee.salary > 150000)
     ScheduleForEarlyRetirement(employee);
```

This might read better with an explanatory variable:

```
boolean isEligibleForEarlyRetirement = employee.age > 60 &&
                                       employee.salary > 150000
if (isEligibleForEarlyRetirement)
     ScheduleForEarlyRetirement(employee);
```

Extract Field（字段抽取）

This refactoring can have a profoundly positive effect. I don't use it often, but when I do, it puts the code on a path to substantial improvement.

It all begins with a failed **Extract Method.** Consider the following class, which converts a CSV file of data into a report. It's a bit of a mess.

```
public class NewCasesReporter {
  public String makeReport(String countyCsv) {
    int totalCases = 0;
    Map<String, Integer> stateCounts = new HashMap<>();
    List<County> counties = new ArrayList<>();

    String[] lines = countyCsv.split("\n");
    for (String line : lines) {
      String[] tokens = line.split(",");
      County county = new County();
      county.county = tokens[0].trim();
      county.state = tokens[1].trim();
      //compute rolling average
      int lastDay = tokens.length - 1;
      int firstDay = lastDay - 7 + 1;
```

8. Kent Beck, *Smalltalk Best Practice Patterns* (Addison-Wesley, 1997), 108.

```
    if (firstDay < 2)
      firstDay = 2;
    double n = lastDay - firstDay + 1;
    int sum = 0;
    for (int day = firstDay; day <= lastDay; day++)
      sum += Integer.parseInt(tokens[day].trim());
    county.rollingAverage = (sum / n);

    //compute sum of cases.
    int cases = 0;
    for (int i = 2; i < tokens.length; i++)
      cases += (Integer.parseInt(tokens[i].trim()));
    totalCases += cases;
    int stateCount = stateCounts.getOrDefault(county.state, 0);
    stateCounts.put(county.state, stateCount + cases);
    counties.add(county);
  }
  StringBuilder report = new StringBuilder("" +
    "County     State     Avg New Cases\n" +
    "======     =====     =============\n");
  for (County county : counties) {
    report.append(String.format("%-11s%-10s%.2f\n",
      county.county,
      county.state,
      county.rollingAverage));
  }
  report.append("\n");
  TreeSet<String> states = new TreeSet<>(stateCounts.keySet());
  for (String state : states)
    report.append(String.format("%s cases: %d\n",
      state, stateCounts.get(state)));
  report.append(String.format("Total Cases: %d\n", totalCases));
  return report.toString();
}

public static class County {
  public String county = null;
  public String state = null;
  public double rollingAverage = Double.NaN;
```

```
  }
}
```

Fortunately for us, the author was kind enough to have written some tests. These tests aren't great, but they'll do.

```
public class NewCasesReporterTest {
  private final double DELTA = 0.0001;
  private NewCasesReporter reporter;

  @Before
  public void setUp() throws Exception {
    reporter = new NewCasesReporter();
  }

  @Test
  public void countyReport() throws Exception {
    String report = reporter.makeReport("" +
      "c1, s1, 1, 1, 1, 1, 1, 1, 1, 7\n" +
      "c2, s2, 2, 2, 2, 2, 2, 2, 2, 7");
    assertEquals("" +
        "County     State     Avg New Cases\n" +
        "======     =====     =============\n" +
        "c1         s1        1.86\n" +
        "c2         s2        2.71\n\n" +
        "s1 cases: 14\n" +
        "s2 cases: 21\n" +
        "Total Cases: 35\n",
      report);
  }

  @Test
  public void stateWithTwoCounties() throws Exception {
    String report = reporter.makeReport("" +
      "c1, s1, 1, 1, 1, 1, 1, 1, 1, 7\n" +
      "c2, s1, 2, 2, 2, 2, 2, 2, 2, 7");
    assertEquals("" +
        "County     State     Avg New Cases\n" +
        "======     =====     =============\n" +
```

```
        "c1          s1          1.86\n" +
        "c2          s1          2.71\n\n" +
        "s1 cases: 35\n" +
        "Total Cases: 35\n",
      report);
  }

  @Test
  public void statesWithShortLines() throws Exception {
    String report = reporter.makeReport("" +
      "c1, s1, 1, 1, 1, 1, 7\n" +
      "c2, s2, 7\n");
    assertEquals("" +
        "County      State       Avg New Cases\n" +
        "======      =====       =============\n" +
        "c1          s1          2.20\n" +
        "c2          s2          7.00\n\n" +
        "s1 cases: 11\n" +
        "s2 cases: 7\n" +
        "Total Cases: 18\n",
      report);
  }
}
```

The tests give us a good idea of what the program is doing. The input is a CSV string. Each line represents a county and has a list of the number of new COVID cases per day. The output is a report that shows the seven-day rolling average of new cases per county and provides some totals for each state, along with a grand total.

Clearly, we want to start extracting methods from this big horrible function. Let's begin with that loop up at the top. That loop does all the math for all the counties, so we should probably call it something like `calculateCounties`.

However, selecting that loop and trying to extract a method produces the dialog shown in Figure 5.1.

Figure 5.1 Extract Method dialog

The IDE wants to name the function `getTotalCases`. You've got to hand it to the IDE authors—they worked pretty hard to try to suggest names. The IDE decided on that name because the code after the loop needs the number of new cases and has no way to get it if this new function doesn't return it.

But we don't want to call the function `getTotalCases`. That's not our intent for this function. We want to call it `calculateCounties`. Moreover, we don't want to pass in those four arguments either. All we really want to pass in to the extracted function is the `lines` array.

So let's hit Cancel and look again.

To refactor this properly, we need to extract some of the local variables within that loop into fields of the surrounding class. We use the **Extract Field** refactoring to do this:

```
public class NewCasesReporter {
  private int totalCases;
  private final Map<String, Integer> stateCounts = new HashMap<>();
  private final List<County> counties = new ArrayList<>();

  public String makeReport(String countyCsv) {
    totalCases = 0;
    stateCounts.clear();
    counties.clear();

    String[] lines = countyCsv.split("\n");
    for (String line : lines) {
      String[] tokens = line.split(",");
      County county = new County();
```

Note that we initialize the values of those variables at the top of the `makeReport` function. This preserves the original behavior.

Now we can extract out the loop without passing in any more variables than we want and without returning the `totalCases`:

```
public class NewCasesReporter {
  private int totalCases;
  private final Map<String, Integer> stateCounts = new HashMap<>();
  private final List<County> counties = new ArrayList<>();

  public String makeReport(String countyCsv) {
    String[] countyLines = countyCsv.split("\n");
    calculateCounties(countyLines);

    StringBuilder report = new StringBuilder("" +
      "County     State     Avg New Cases\n" +
      "======     =====     =============\n");
    for (County county : counties) {
```

```
    report.append(String.format("%-11s%-10s%.2f\n",
      county.county,
      county.state,
      county.rollingAverage));
  }
  report.append("\n");
  TreeSet<String> states = new TreeSet<>(stateCounts.keySet());
  for (String state : states)
    report.append(String.format("%s cases: %d\n",
      state, stateCounts.get(state)));
  report.append(String.format("Total Cases: %d\n", totalCases));
  return report.toString();
}

private void calculateCounties(String[] lines) {
  totalCases = 0;
  stateCounts.clear();
  counties.clear();

  for (String line : lines) {
    String[] tokens = line.split(",");
    County county = new County();
    county.county = tokens[0].trim();
    county.state = tokens[1].trim();
    //compute rolling average
    int lastDay = tokens.length - 1;
    int firstDay = lastDay - 7 + 1;
    if (firstDay < 2)
      firstDay = 2;
    double n = lastDay - firstDay + 1;
    int sum = 0;
    for (int day = firstDay; day <= lastDay; day++)
      sum += Integer.parseInt(tokens[day].trim());
    county.rollingAverage = (sum / n);

    //compute sum of cases.
    int cases = 0;
    for (int i = 2; i < tokens.length; i++)
      cases += (Integer.parseInt(tokens[i].trim()));
```

```
      totalCases += cases;
      int stateCount = stateCounts.getOrDefault(county.state, 0);
      stateCounts.put(county.state, stateCount + cases);
      counties.add(county);
    }
  }

  public static class County {
    public String county = null;
    public String state = null;
    public double rollingAverage = Double.NaN;
  }
}
```

Now, with those variables as fields, we can continue to extract and rename to our heart's delight.

```
public class NewCasesReporter {
  private int totalCases;
  private final Map<String, Integer> stateCounts = new HashMap<>();
  private final List<County> counties = new ArrayList<>();

  public String makeReport(String countyCsv) {
    String[] countyLines = countyCsv.split("\n");
    calculateCounties(countyLines);

    StringBuilder report = makeHeader();
    report.append(makeCountyDetails());
    report.append("\n");
    report.append(makeStateTotals());
    report.append(String.format("Total Cases: %d\n", totalCases));
    return report.toString();
  }

  private void calculateCounties(String[] countyLines) {
    totalCases = 0;
    stateCounts.clear();
    counties.clear();
```

```
    for (String countyLine : countyLines)
      counties.add(calcluateCounty(countyLine));
  }

  private County calcluateCounty(String line) {
    County county = new County();
    String[] tokens = line.split(",");
    county.county = tokens[0].trim();
    county.state = tokens[1].trim();

    county.rollingAverage = calculateRollingAverage(tokens);

    int cases = calculateSumOfCases(tokens);
    totalCases += cases;
    incrementStateCounter(county.state, cases);

    return county;
  }

  private double calculateRollingAverage(String[] tokens) {
    int lastDay = tokens.length - 1;
    int firstDay = lastDay - 7 + 1;
    if (firstDay < 2)
      firstDay = 2;
    double n = lastDay - firstDay + 1;
    int sum = 0;
    for (int day = firstDay; day <= lastDay; day++)
      sum += Integer.parseInt(tokens[day].trim());
    return (sum / n);
  }

  private int calculateSumOfCases(String[] tokens) {
    int cases = 0;
    for (int i = 2; i < tokens.length; i++)
      cases += (Integer.parseInt(tokens[i].trim()));
    return cases;
  }

  private void incrementStateCounter(String state, int cases) {
```

```
    int stateCount = stateCounts.getOrDefault(state, 0);
    stateCounts.put(state, stateCount + cases);
  }

  private StringBuilder makeHeader() {
    return new StringBuilder("" +
      "County     State     Avg New Cases\n" +
      "======     =====     =============\n");
  }

  private StringBuilder makeCountyDetails() {
    StringBuilder countyDetails = new StringBuilder();
    for (County county : counties) {
      countyDetails.append(String.format("%-11s%-10s%.2f\n",
        county.county,
        county.state,
        county.rollingAverage));
    }
    return countyDetails;
  }

  private StringBuilder makeStateTotals() {
    StringBuilder stateTotals = new StringBuilder();
    TreeSet<String> states = new TreeSet<>(stateCounts.keySet());
    for (String state : states)
      stateTotals.append(String.format("%s cases: %d\n",
        state, stateCounts.get(state)));
    return stateTotals;
  }

  public static class County {
    public String county = null;
    public String state = null;
    public double rollingAverage = Double.NaN;
  }
}
```

This is much better, but I don't like the fact that the code that formats the report is in the same class with the code that calculates the data. That's a violation of the single responsibility principle because the format of the report and the calculations are very likely to change for different reasons.

In order to pull the calculation portion of the code out into a new class, we use the **Extract Superclass** refactoring to pull the calculations up into a superclass named `NewCasesCalculator`. `NewCasesReporter` will derive from it.

```
public class NewCasesCalculator {
  protected final Map<String, Integer> stateCounts = new HashMap<>();
  protected final List<County> counties = new ArrayList<>();
  protected int totalCases;

  protected void calculateCounties(String[] countyLines) {
    totalCases = 0;
    stateCounts.clear();
    counties.clear();

    for (String countyLine : countyLines)
      counties.add(calcluateCounty(countyLine));
  }

  private County calcluateCounty(String line) {
    County county = new County();
    String[] tokens = line.split(",");
    county.county = tokens[0].trim();
    county.state = tokens[1].trim();

    county.rollingAverage = calculateRollingAverage(tokens);

    int cases = calculateSumOfCases(tokens);
    totalCases += cases;
    incrementStateCounter(county.state, cases);

    return county;
  }

  private double calculateRollingAverage(String[] tokens) {
```

```
    int lastDay = tokens.length - 1;
    int firstDay = lastDay - 7 + 1;
    if (firstDay < 2)
      firstDay = 2;
    double n = lastDay - firstDay + 1;
    int sum = 0;
    for (int day = firstDay; day <= lastDay; day++)
      sum += Integer.parseInt(tokens[day].trim());
    return (sum / n);
  }

  private int calculateSumOfCases(String[] tokens) {
    int cases = 0;
    for (int i = 2; i < tokens.length; i++)
      cases += (Integer.parseInt(tokens[i].trim()));
    return cases;
  }

  private void incrementStateCounter(String state, int cases) {
    int stateCount = stateCounts.getOrDefault(state, 0);
    stateCounts.put(state, stateCount + cases);
  }

  public static class County {
    public String county = null;
    public String state = null;
    public double rollingAverage = Double.NaN;
  }
}

=======

public class NewCasesReporter extends NewCasesCalculator {
  public String makeReport(String countyCsv) {
    String[] countyLines = countyCsv.split("\n");
    calculateCounties(countyLines);

    StringBuilder report = makeHeader();
    report.append(makeCountyDetails());
```

```
    report.append("\n");
    report.append(makeStateTotals());
    report.append(String.format("Total Cases: %d\n", totalCases));
    return report.toString();
  }

  private StringBuilder makeHeader() {
    return new StringBuilder("" +
      "County     State     Avg New Cases\n" +
      "======     =====     =============\n");
  }

  private StringBuilder makeCountyDetails() {
    StringBuilder countyDetails = new StringBuilder();
    for (County county : counties) {
      countyDetails.append(String.format("%-11s%-10s%.2f\n",
        county.county,
        county.state,
        county.rollingAverage));
    }
    return countyDetails;
  }

  private StringBuilder makeStateTotals() {
    StringBuilder stateTotals = new StringBuilder();
    TreeSet<String> states = new TreeSet<>(stateCounts.keySet());
    for (String state : states)
      stateTotals.append(String.format("%s cases: %d\n",
        state, stateCounts.get(state)));
    return stateTotals;
  }
}
```

This partitioning separates things out very nicely. Reporting and calculation are accomplished in separate modules. And all because of that initial **Extract Field**.

Rubik's Cube（魔方）

So far, I've tried to show you how powerful a small set of refactorings can be. In my normal work, I seldom use more than the ones I've shown you. The trick is to learn them well and understand all the details of the IDE and the tricks for using them.

I have often compared refactoring to solving a Rubik's cube. If you've never solved one of these puzzles, it would be worth your time to learn how. Once you know the trick, it's relatively easy.

It turns out that there are a set of "operations" that you can apply to the cube that preserve most of the cube's positions but change certain positions in predictable ways. Once you know three or four of those operations, you can incrementally manipulate the cube into a solvable position.

The more operations you know and the more adept you are at performing them, the faster and more directly you can solve the cube. But you'd better learn those operations well. One missed step and the cube melts down into a random distribution of cubies, and you have to start all over.

Refactoring code is a lot like this. The more refactorings you know and the more adept you are at using them, the easier it is to push, pull, and stretch the code in any direction you desire.

Oh, and you'd better have tests too. Without them, meltdowns are a near certainty.

The Disciplines（纪律）

Refactoring is safe, easy, and powerful if you approach it in a regular and disciplined manner. If, on the other hand, you approach it as an ad hoc, temporary, and sporadic activity, that safety and power can quickly evaporate.

TESTS（测试）

The first of the disciplines, of course, is tests. Tests, tests, tests, tests, and more tests. To safely and reliably refactor your code, you need a test suite that you trust with your life. You need tests.

QUICK TESTS（快速测试）

The tests also need to be quick. Refactoring just doesn't work well if your tests take hours (or even minutes) to run.

In large systems, no matter how hard you try to reduce test time, it's hard to reduce it to less than a few minutes. For this reason, I like to organize my test suite such that I can quickly and easily run *the relevant subset* of tests that check the part of the code I am refactoring at the moment. This usually allows me to reduce the test time from minutes to sub-seconds. I run the whole suite once every hour or so just to make sure no bugs leaked out.

BREAK DEEP ONE-TO-ONE CORRESPONDENCES（打破紧密的一一对应关系）

Creating a test structure that allows relevant subsets to be run means that, at the level of modules and components, the design of your tests will mirror the design of your code. There will likely be a one-to-one correspondence between your high-level test modules and your high-level production code modules.

As we learned in the previous section, deep one-to-one correspondences between tests and code lead to fragile tests.

The speed benefit of being able to run relevant subsets of tests is much greater than the cost of one-to-one coupling *at that level*. But in order to prevent fragile tests, we don't want the one-to-one correspondence to continue. So, below the level of modules and components, we purposely break that one-to-one correspondence.

REFACTOR CONTINUOUSLY（持续重构）

When I cook a meal, I make it a rule to clean the preparation dishes as I proceed.[9] I do not let them pile up in the sink. There's always enough time to clean the used utensils and pans while the food is cooking.

Refactoring is like that too. Don't wait to refactor. Refactor as you go. Keep the red → green → refactor loop in your mind, and spin around that loop every few minutes. That way, you will prevent the mess from building so large that it starts to intimidate you.

REFACTOR MERCILESSLY（果断重构）

Merciless refactoring was one of Kent Beck's sound bites for Extreme Programming. It was a good one. The discipline is simply to be courageous when you refactor. Don't be afraid to try things. Don't be reluctant to make changes. Manipulate the code as though it is clay and you are the sculptor. Fear of the code is the mind-killer, the dark path. Once you start down the dark path, forever will it dominate your destiny. Consume you it will.

KEEP THE TESTS PASSING!（让测试始终能通过）

Sometimes you will realize that you've made a structural error and that a large swath of code needs to change. This can happen when a new requirement that invalidates your current design comes along. It can also happen out of the blue when, one day, you suddenly realize that there's a better structure for the future of your project.

You must be merciless, but you must also be smart. Never break the tests! Or, rather, never leave them broken for more than a few minutes at a time.

If the restructuring is going to take hours or days to complete, then do the restructuring in small chunks while you keep everything passing and while you continue to do other activities.

---

9. My wife disputes this claim.

For example, let's say that you realize that you need to change a fundamental data structure in the system—a data structure that large swathes of the code use. If you were to change that data structure, those swaths would stop working and many tests would break.

Instead, you should create a new data structure that mirrors the content of the old data structure. Then, gradually, move each portion of the code from the old data structure to the new data structure while keeping the tests passing.

While this is going on, you may also be adding new features and fixing bugs according to your regular schedule of work. There is no need to ask for special time to perform this restructuring. You can keep on doing other work while you opportunistically manipulate the code until the old data structure is no longer used and can be deleted.

This may take weeks or even months, depending on how significant the restructuring is. Even so, at no time would the system be down for deployment. Even while the restructuring is only partially complete, the tests still pass and the system can be deployed into production.

Leave Yourself an Out（留条出路）

When flying into an area where the weather might not be so good, pilots are taught to always make sure they leave an avenue of escape. Refactoring can be a bit like that. Sometimes you start a series of refactorings that, after an hour or two, leads you to a dead end. The idea you started with just didn't pan out for some reason.

In situations like this, `git reset --hard` can be your friend.

So, when beginning such a sequence of refactorings, make sure to tag your source repository so you can back out if you need to.

Conclusion（小结）

I kept this chapter intentionally brief because there were only a few ideas that I wanted to add to Martin Fowler's *Refactoring*. Again, I urge you to refer to that book for an in-depth understanding.

The best approach to refactoring is to develop a comfortable repertoire of refactorings that you use frequently and to have a good working knowledge of many others. If you use an IDE that provides refactoring operations, make sure you understand them in detail.

Refactoring makes no sense without tests. Without tests, the opportunities for error are just too common. Even the automated refactorings that your IDE provides can sometimes make mistakes. So *always* back up your refactoring efforts with a comprehensive suite of tests.

Finally, be disciplined. Refactor frequently. Refactor mercilessly. And refactor without apology. Never, ever ask permission to refactor.

6 Simple Design

（简单设计）

Design. The Holy Grail and ultimate objective of the software craft. We all seek to create a design so perfect that features can be added without effort and without fuss. A design so robust that, despite months and years of constant maintenance, the system remains facile and flexible. Design, in the end, is what it's all about.

I have written a great deal about design. I have written books about design principles, design patterns, and architecture. And I am far from the only author to focus on this topic. The amount of literature on software design is enormous.

But that's not what this chapter is about. You would be well advised to research the topic of design, read those authors, and understand the principles, the patterns, and the overall gestalt of software design and architecture.

But the key to it all, the aspect of design that imbues it with all the characteristics that we desire, is—in a word—*simplicity*. As Chet Hendrickson[1] once said, "Uncle Bob wrote thousands of pages on clean code. Kent Beck wrote four lines." It is those four lines we focus on here.

It should be obvious, on the face of it, that the best design for a system is the simplest design that supports all the required features of that system while simultaneously affording the greatest flexibility for change. However, that leaves us to ponder the meaning of simplicity.[2] Simple does not mean easy. Simple means untangled, and untangling things is *hard*.

What things get tangled in software systems? The most expensive and significant entanglements are those that convolve high-level policies with low-level details. You create terrible complexities when you conjoin SQL with HTML, or frameworks with core values, or the format of a report with the business rules that calculate the reported values. These entanglements are

1. As cited by Martin Fowler in a tweet quoting Chet Hendrickson at AATC2017. I was in attendance when Chet said it, and I completely agreed with him.
2. In 2012, Rich Hickey gave a wonderful talk, *Simple Made Easy*. I encourage you to listen to it. https://www.youtube.com/watch?v=oytL881p-nQ.

easy to write, but they make it hard to add new features, hard to fix bugs, and hard to improve and clean the design.

A simple design is a design in which high-level policies are ignorant of low-level details. Those high-level policies are sequestered and isolated from low-level details such that changes to the low-level details have no impact on the high-level policies.[3]

The primary means for creating this separation and isolation is *abstraction*. Abstraction is the amplification of the essential and the elimination of the irrelevant. High-level policies are essential, so they are amplified. Low-level details are irrelevant, so they are isolated and sequestered.

The physical means we employ for this abstraction is *polymorphism*. We arrange high-level policies to use polymorphic interfaces to manage the low-level details. Then we arrange the low-level details as implementations of those polymorphic interfaces. This practice keeps all source code dependencies pointing from low-level details to high-level policies and keeps high-level policies ignorant of the implementations of the low-level details. Low-level details can be changed without affecting the high-level policies (Figure 6.1).

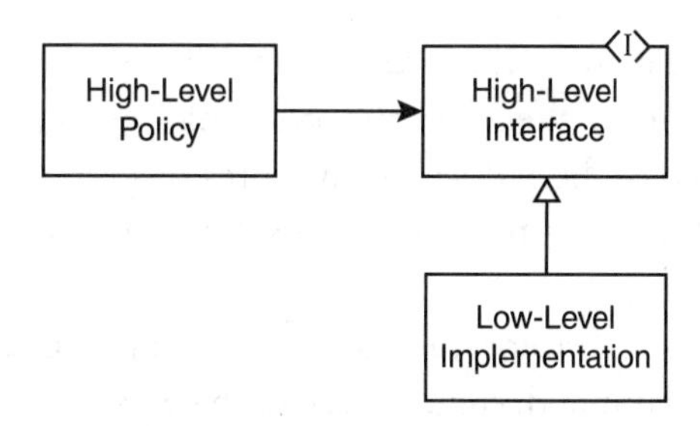

Figure 6.1 Polymorphism

3. I write a great deal about this in *Clean Architecture: A Craftsman's Guide to Software Structure and Design* (Addison-Wesley, 2018).

If the best design of a system is the simplest design that supports the features, then we can say that such a design must have the fewest abstractions that manage to isolate high-level policy from low-level detail.

And yet, this was precisely the opposite of the strategy we employed throughout the 1980s and 1990s. In those days, we were obsessed with *future-proofing* our code by *putting the hooks in* for the changes we anticipated in the future.

We took this path because software, in those days, was hard to change—even if the design was simple.

Why was software hard to change? Because build times were long and test times were longer.

In the 1980s, a small system might require an hour or more to build and many hours to test. Tests, of course, were manual and were therefore woefully inadequate. As a system grew larger and more complicated, the programmers became ever-more afraid of making changes. This led to a mentality of overdesign, which drove us to create systems that were far more complicated than necessary for the features they had.

We reversed course in the late 1990s with the advent of Extreme Programming and then Agile. By then, our machines had become so massively powerful that build times could be reduced to minutes or even seconds, and we found that we could afford to automate tests that could run very quickly.

Driven by this technological leap, the discipline of YAGNI and the four principles of simple design, described by Kent Beck, became practicable.

YAGNI

What if you aren't gonna need it?

In 1999, I was teaching an Extreme Programming course with Martin Fowler, Kent Beck, Ron Jeffries, and a host of others. The topic turned to the dangers of overdesign and premature generalization. Someone wrote YAGNI on the whiteboard and said, "You aren't gonna need it." Beck interrupted and said something to the effect maybe you are gonna need it, but you should ask yourself, "What if you aren't?"

That was the original question that YAGNI asked. Every time you thought to yourself, *I'm going to need this hook,* you then asked yourself what would happen if you didn't put the hook in. If the cost of leaving the hook out was tolerable, then you probably shouldn't put it in. If the cost of carrying the hook in the design, year after year, would be high but the odds that you'd eventually need that hook were low, you probably shouldn't put that hook in.

It is hard to imagine the furor this new perspective raised in the late 1990s. Designers everywhere had gotten very used to putting all the hooks in. In those days, putting the hooks in was considered conventional wisdom and a best practice.

So, when the YAGNI discipline of Extreme Programming surfaced, it was roundly criticized and panned as heresy and claptrap.

Nowadays, ironically, it is one of the most important disciplines of good software design. If you have a good suite of tests and you are skilled at the discipline of refactoring, then the cost of adding a new feature and updating the design to support that new feature will almost certainly be smaller than the cost of implementing and maintaining all the hooks you might need one day.

Hooks are problematic in any case. We seldom get them right. That's because we are not particularly good at predicting what customers will actually do. Consequently, we tend to put in far more hooks than we need and base those hooks on assumptions that rarely pan out.

The bottom line is that that the effect that gigahertz clock rates and terabyte memories had on the process of software design and architecture took us all by surprise. We did not realize until the late 1990s that those advances would allow us to vastly *simplify* our designs.

It is one of the great ironies of our industry that the exponential increase of Moore's law that drove us to build ever-more complex software systems also made it possible to simplify the designs of those systems.

YAGNI, it turns out, is the unintended consequence of the virtually limitless computer power now at our disposal. Because our build times have shrunk down into the seconds and because we can afford to write and execute comprehensive test suites that execute in seconds, we can afford to *not* put the hooks in and instead refactor the designs as the requirements change.

Does this mean that we never put the hooks in? Do we always design our systems only for the features we need today? Do we never think ahead and plan for the future?

No, that's not what YAGNI means. There are times when putting a particular hook in is a good idea. Future-proofing the code is not dead, and it is always wise to think of the future.

It's just that the trade-offs have changed so dramatically in the last few decades that it is now usually better to leave the majority of the hooks out. And that's why we ask the question:

What if you aren't gonna need it?

Covered by Tests（用测试覆盖）

The first time I ran across Beck's rules of simple design was in the first edition of *Extreme Programming Explained*.[4] At that time, the four rules were as follows:

1. The system (code and tests) must communicate everything you want to communicate.
2. The system must contain no duplicate code.

4. Kent Beck, *Extreme Programming Explained* (Addison-Wesley, 1999).

3. The system should have the fewest possible classes.
4. The system should have the fewest possible methods.

By 2011, they had evolved to these:

1. Tests pass.
2. Reveal intent.
3. No duplication.
4. Small.

By 2014, Corey Haines had written a book[5] about those four rules:

In 2015, Martin Fowler wrote a blog[6] about them in which he rephrased them:

1. Passes the tests.
2. Reveals intention.
3. No duplication.
4. Fewest elements.

In this book I express the first rule as

1. *Covered by tests.*

Notice how the emphasis of that first rule has changed over the years. The first rule split in two and the last two rules merged into one. Notice also that, as the years went by, tests grew in importance from communication to coverage.

5. Corey Haines, *Understanding the Four Rules of Simple Design* (Leanpub, 2014).
6. Martin Fowler, "BeckDesignRules," March 2, 2015, https://martinfowler.com/bliki/BeckDesignRules.html.

Coverage（覆盖）

The concept of tests coverage is an old one. The first mention I was able to find goes all the way back to 1963.[7] The article begins with two paragraphs that I think you'll find interesting, if not evocative.

> *Effective program checkout is imperative to any complex computer program. One or more test cases are always run for a program before it is considered ready for application to the actual problem. Each test case checks that portion of the program actually used in its computation. Too often, however, mistakes show up as late as several months (or even years) after a program has been put into operation. This is an indication that the portions of the program called upon only by rarely occurring input conditions have not been properly tested during the checkout stage.*
>
> *In order to rely with confidence upon any particular program it is not sufficient to know that the program works most of the time or even that it has never made a mistake so far. The real question is whether it can be counted upon to fulfill its functional specifications successfully every single time. This means that, after a program has passed the checkout stage, there should be no possibility that an unusual combination of input data or conditions may bring to light an unexpected mistake in the program. Every portion of the program must be utilized during checkout in order that its correctness be confirmed.*

Nineteen sixty-three was only seventeen years after the very first program ran on the very first electronic computer,[8] and already we knew that the only way to effectively mitigate the threat of software errors is to test every single line of code.

Code coverage tools have been around for decades. I don't remember when I first encountered them. I think it was in the late 1980s or early 1990s. At the time, I was working on Sun Microsystems Sparc Stations, and Sun had a tool called `tcov`.

---

7. Joan Miller and Clifford J Maloney, "Systematic Mistake Analysis of Digital Computer Programs," *Communications of the ACM* 6, no. 2 (1963): 58–63.
8. Presuming that the first computer was the Automated Computing Engine and that the first program executed in 1946.

I don't remember when I first heard the question, What's your code coverage? It was probably in the very early 2000s. But thereafter the notion that code coverage was a number became pretty much universal.

Since then, it has become relatively commonplace for software teams to run a code coverage tool as part of their continuous build process and to publish the code coverage number for each build.

What is a good code coverage number? Eighty percent? Ninety percent? Many teams are more than happy to report such numbers. But six decades before the publication of this book, Miller and Maloney answered the question very differently: Their answer was 100 percent.

What other number could possibly make sense? If you are happy with 80 percent coverage, it means you don't know if 20 percent of your code works. How could you possibly be happy with that? How could your customers be happy with that?

So, when I use the term *covered* in the first rule of simple design, I mean *covered*. I mean 100 percent line coverage and 100 percent branch coverage.

An Asymptotic Goal（渐近目标）

You might complain that 100 percent is an unreachable goal. I might even agree with you. Achieving 100 percent line and branch coverage is no mean feat. It may, in fact, be impractical depending on the situation. But that does not mean that your coverage cannot be improved.

Think of the number 100 percent as an asymptotic goal. You may never reach it, but that's no excuse for not trying to get closer and closer with every check-in.

I have personally participated in projects that grew to many tens of thousands of lines of code while constantly keeping the code coverage in the very high nineties.

DESIGN?（设计？）

But what does high code coverage have to do with simple design? Why is coverage the first rule?

> *Testable code is decoupled code.*

In order to achieve high line and branch coverage of each individual part of the code, each of those parts must be made accessible to the test code. That means those parts must be so well decoupled from the rest of the code that they can be isolated and invoked from an individual test. Therefore, those tests not only are tests of behavior but also are tests of decoupling. The act of writing isolated tests is an act of design, because the code being tested must be designed to be tested.

In Chapter 4, "Testing Design," we talked about how the test code and the production code evolve in different directions in order to keep the tests from coupling too strongly to the production code. This prevents the problem of fragile tests. But the problem of fragile tests is no different from the problem of fragile modules, and the cure for both is the same. If the design of your system keeps your tests from being fragile, it will also keep the other elements of your system from being fragile.

BUT THERE'S MORE（但还有更多好处）

Tests don't just drive you to create decoupled and robust designs. They also allow you to improve those designs with time. As we have discussed many times before in these pages, a trusted suite of tests vastly reduces the fear of change. If you have such a suite, and if that suite executes quickly, then you can improve the design of the code every time you find a better approach. When the requirements change in a way that the current design does not easily accommodate, the tests will allow you to fearlessly shift the design to better match those new requirements.

And this is why this rule is the first and most important rule of simple design. Without a suite of tests that covers the system, the other three rules become impractical, because those rules are best applied *after the fact*. Those other three rules are rules that involve refactoring. And refactoring is virtually impossible without a good, comprehensive suite of tests.

Maximize Expression（充分表达）

In the early decades of programming, the code we wrote could not reveal intent. Indeed, the very name "code" suggests that intent is obscured. Back in those days, code looked like what is shown in Figure 6.2.

```
------

/ROUTINE TO TYPE A MESSAGE                    PAL8-V10D NO DATE   PAGE 1

              /ROUTINE TO TYPE A MESSAGE
        0200            *200
        7600            MONADR=7600
00200   7300   START,   CLA CLL           /CLEAR ACCUMULATOR AND LINK
00201   6046            TLS               /CLEAR TERMINAL FLAG
00202   1216            TAD BUFADR        /SET UP POINTER
00203   3217            DCA PNTR          /FOR GETTING CHARACTERS
00204   6041   NEXT,    TSF               /SKIP IF TERMINAL FLAG SET
00205   5204            JMP .-1           /NO: CHECK AGAIN
00206   1617            TAD I PNTR        /GET A CHARACTER
00207   6046            TLS               /PRINT A CHARACTER
00210   2217            ISZ PNTR          /DONE YET?
00211   7300            CLA CLL           /CLEAR ACCUMULATOR AND LINK
00212   1617            TAD I PNTR        /GET ANOTHER CHARACTER
00213   7640            SZA CLA           /JUMP ON ZERO AND CLEAR
00214   5204            JMP NEXT          /GET READY TO PRINT ANOTHER
00215   5631            JMP I MON         /RETURN TO MONITOR
00216   0220   BUFADR,  BUFF              /BUFFER ADDRESS
00217   0220   PNTR,    BUFF              /POINTER
00220   0215   BUFF,    215;212;"H;"E;"L;"L;"O;"!;0
00221   0212
00222   0310
00223   0305
00224   0314
00225   0314
00226   0317
00227   0241
00230   0000
00231   7600   MON,     MONADR            /MONITOR ENTRY POINT
```

Figure 6.2 An example of an early program

Notice the ubiquitous comments. These were absolutely necessary because the code itself revealed nothing at all about the intent of the program.

However, we no longer work in the 1970s. The languages that we use are *immensely* expressive. With the proper discipline, we can produce code that reads like "well written-prose [that] never obscures the designer's intent."[9]

As an example of such code, consider this little bit of Java from the video store example in Chapter 4:

```
public class RentalCalculator {
  private List<Rental> rentals = new ArrayList<>();

  public void addRental(String title, int days) {
    rentals.add(new Rental(title, days));
  }

  public int getRentalFee() {
    int fee = 0;
    for (Rental rental : rentals)
      fee += rental.getFee();
    return fee;
  }

  public int getRenterPoints() {
    int points = 0;
    for (Rental rental : rentals)
      points += rental.getPoints();
    return points;
  }
}
```

If you were not a programmer on this project, you might not understand everything that is going on in this code. However, after even the most cursory glance, the basic intent of the designer is easy to identify. The names of the variables, functions, and types are deeply descriptive. The structure of the algorithm is easy to see. This code is expressive. This code is simple.

9. Martin, *Clean Code*, p. 8 (personal correspondence with Grady Booch).

The Underlying Abstraction（底层抽象）

Lest you think that expressivity is solely a matter of nice names for functions and variables, I should point out that there is another concern: the separation of levels and the exposition of the underlying abstraction.

A software system is expressive if each line of code, each function, and each module lives in a well-defined partition that clearly depicts the level of the code and its place in the overall abstraction.

You may have found that last sentence difficult to parse, so let me be a bit clearer by being much more longwinded.

Imagine an application that has a complex set of requirements. The example I like to use is a payroll system.

- Hourly employees are paid every Friday on the basis of the timecards they have submitted. They are paid time and a half for every hour they work after forty hours in a week.
- Commissioned employees are paid on the first and third Friday of every month. They are paid a base salary plus a commission on the sales receipts they have submitted.
- Salaried employees are paid on the last day of the month. They are paid a fixed monthly salary.

It should not be hard for you to imagine a set of functions with a complex `switch` statement or `if/else` chain that captures these requirements. However, such a set of functions is likely to obscure the underlying abstraction. What is that underlying abstraction?

```
public List<Paycheck> run(Database db) {
  Calendar now = SystemTime.getCurrentDate();
  List<Paycheck> paychecks = new ArrayList<>();
  for (Employee e : db.getAllEmployees()) {
    if (e.isPayDay(now))
      paychecks.add(e.calculatePay());
```

```
    }
    return paychecks;
  }
```

Notice that there is no mention of any of the hideous details that dominate the requirements. The underlying truth of this application is that we need to pay all employees on their payday. Separating the high-level policy from the low-level detail is the most fundamental part of making a design simple and expressive.

Tests: The Other Half of the Problem（再论测试：问题的后半部分）

Look back at Beck's original first rule:

> *1. The system (code and tests) must communicate everything you want to communicate.*

There is a reason he phrased it that way, and in some ways, it is unfortunate that the phrasing was changed.

No matter how expressive you make the production code, it cannot communicate the context in which it is used. That's the job of the tests.

Every test you write, especially if those tests are isolated and decoupled, is a demonstration of how the production code is intended to be used. Well-written tests are example use cases for the parts of the code that they test.

Thus, *taken together,* the code and the tests are an expression of what each element of the system does and how each element of the system should be used.

What does this have to do with design? Everything, of course. Because the primary goal we wish to achieve with our designs is to make it easy for other programmers to understand, improve, and upgrade our systems. And there is no better way to achieve that goal than to make the system express what it does and how it is intended to be used.

Minimize Duplication（尽量减少重复）

In the early days of software, we had no source code editors at all. We wrote our code, using #2 pencils, on preprinted coding forms. The best editing tool we had was an eraser. We had no practical means to copy and paste.

Because of that, we did not duplicate code. It was easier for us to create a single instance of a code snippet and put it into a subroutine.

But then came the source code editors, and with those editors came copy/paste operations. Suddenly it was much easier to copy a snippet of code and paste it into a new location and then fiddle with it until it worked.

Thus, as the years went by, more and more systems exhibited massive amounts of duplication in the code.

Duplication is usually problematic. Two or more similar stretches of code will often need to be modified together. Finding those similar stretches is hard. Properly modifying them is even harder because they exist in different contexts. Thus, duplication leads to fragility.

In general, it is best to reduce similar stretches of code into a single instance by abstracting the code into a new function and providing it with appropriate arguments that communicate any differences in context.

Sometimes that strategy is not workable. For example, sometimes the duplication is in code that traverses a complex data structure. Many different parts of the system may wish to traverse that structure and will use the same looping and traversal code only to then operate on the data structure in the body of that code.

As the structure of the data changes over time, programmers will have to find all the duplications of the traversal code and properly update them. The more the traversal code is duplicated, the higher the risk of fragility.

The duplications of the traversal code can be eliminated by encapsulating it in once place and using lambdas, *Command* objects, the *Strategy* pattern or even the *Template Method* pattern[10] to pass the necessary operations into the traversal.

Accidental Duplication（意外重复）

Not all duplication should be eliminated. There are instances in which two stretches of code may be very similar, even identical, but will change for very different reasons.[11] I call this *accidental duplication*. Accidental duplicates should not be eliminated. The duplication should be allowed to persist. As the requirements change, the duplicates will evolve separately and the accidental duplication will dissolve.

It should be clear that managing duplication is nontrivial. Identifying which duplications are real and which are accidental, and then encapsulating and isolating the real duplications, requires a significant amount of thought and care.

Determining the real duplications from the accidental duplications depends strongly on how well the code expresses its intent. Accidental duplications have divergent intent. Real duplications have convergent intent.

Encapsulating and isolating the real duplications, using abstraction, lambdas, and design patterns, involves a substantial amount of refactoring. And refactoring requires a good solid test suite.

Therefore, eliminating duplication is third in the priority list of the rules of simple design. First come the tests and the expression.

---

10. Erich Gamma, Richard Helm, Ralph Johnson, and John M Vlissides, *Design Patterns: Elements of Reusable Object-Oriented Software* (Addison-Wesley, 1995).
11. See the single responsibility principle. Robert C. Martin, *Agile Software Development: Principles, Patterns, and Practices* (Pearson, 2003).

Minimize Size（尺寸尽量小）

A simple design is composed of simple elements. Simple elements are small. The last rule of simple design states that for each function you write, after you've gotten all the tests to pass, and after you have made the code as expressive as possible, and after you have minimized duplication, *then* you should work to decrease the size of the code within each function without violating the other three principles.

How do you do that? Mostly by extracting more functions. As we discussed in Chapter 5, "Refactoring," you extract functions until you cannot extract any more.

This practice leaves you with nice small functions that have nice long names that help to make the functions very small and very expressive.

Simple Design（简单设计）

Many years back, Kent Beck and I were having a discussion on the principles of design. He said something that has always stuck with me. He said that if you followed these four things as diligently as possible, all other design principles would be satisfied—the principles of design can be reduced to coverage, expression, singularization, and reduction.

I don't know if this is true or not. I don't know if a perfectly covered, expressed, singularized, and reduced program necessarily conforms to the open-closed principle or the single responsibility principle. What I am very sure of, however, is that knowing and studying the principles of good design and good architecture (e.g., the SOLID principles) makes it much easier to create well-partitioned and simple designs.

This is not a book about those principles. I have written about them many times before,[12] as have others. I encourage you to read those works and study those principles as part of maturing in your craft.

---

12. See Martin, *Clean Code; Clean Architecture;* and *Agile Software Development: Principles, Patterns, and Practices.*

7 Collaborative Programming

（协同编程）

What does it mean to be part of a team? Imagine a team of players working to move the ball down the field against their opponents. Imagine one of those players trips and falls, but the play continues. What do the other players do?

The other players adapt to the new reality by shifting their field positions in order to *keep the ball moving down the field.*

That's how a team behaves. When a team member goes down, the team covers for that member until they are back up on their feet.

How do we make a programming team into a team like that? How can the team cover for someone who gets sick for a week or just has a bad programming day? We collaborate! We work together so that knowledge of the whole system spreads through the team.

When Bob goes down, someone else who has recently worked with Bob can cover the hole until Bob gets back on his feet.

The old adage that two heads are better than one is the basic premise of collaborative programming. When two programmers collaborate, it is often called pair programming.[1] With three or more, it goes by mob programming.[2]

The discipline involves two or more people working together at the same time, on the same code. Nowadays, this is typically done by using screen-sharing software. Both programmers see the same code on their screens. Both can use their mouse and keyboard to manipulate that code. Their workstations are slaved to each other either locally or remotely.

Collaboration like this is not something that should generally be done 100 percent of the time. Rather, collaboration sessions are generally brief, informal, and intermittent. The total time a team should work collaboratively depends on the maturity, skill, geography, and demographics of the team and should be somewhere in the range of 20 to 70 percent.[3]

---

1. Laurie Williams and Robert Kessler, *Pair Programming Illuminated* (Addison-Wesley, 2002).
2. Mark Pearl, *Code with the Wisdom of the Crowd* (Pragmatic Bookshelf, 2018).
3. There are teams who pair 100 percent of the time. They seem to enjoy it, and more power to them.

A collaboration session can last as little as 10 minutes or as long as an hour or two. Sessions that are shorter or longer than those limits are likely to be less than helpful. My favorite collaboration strategy is to use the Pomodoro technique.[4] This technique divides time into "tomatoes" of 20 minutes or so, with short breaks in between. A collaboration session should last between one and three tomatoes.

Collaboration sessions are much shorter lived than programming tasks. Individual programmers take responsibility for particular tasks and then, from time to time, invite collaborators to help meet those responsibilities.

No one person is in charge of a collaboration session or of the code being manipulated within a session. Rather, every participant is an equal author of and contributor to the code under consideration. The programmer responsible for the task is the final arbiter should a dispute arise in the midst of a session.

In a session, all eyes are on the screen, all minds are engaged on the problem. One or two people may be seated at keyboards, but those seats can change frequently within the session. Think of the session as simultaneously being a live coding exercise and a code review.

Collaboration sessions are very intense and require a lot of mental and emotional energy. One or two hours at that level of intensity is likely all that the average programmer can tolerate before needing to break away to something less consuming.

You might worry that collaboration like this is an inefficient use of manpower, that people working independently can get more done than people working together. This does not turn out to be particularly true. Studies[5] of programmers working in pairs have shown that the productivity within a pairing session drops by only about 15 percent, as opposed to the feared

4. Francesco Cirillo, *The Pomodoro Technique* (Currency Publishing, 2018).

5. Two such studies are "Strengthening the Case for Pair Programming" by Laurie Williams, Robert R. Kessler, Ward Cunningham, and Ron Jeffries, *IEEE Software* 17, no. 4 (2000), 19–25; and "The Case for Collaborative Programming" by J. T. Nosek, *Communications of the ACM* 41, no. 3 (1998), 105–108.

50 percent. However, during that pairing session, the pair creates about 15 percent fewer defects and (more important) about 15 percent less code per feature.

Those last two statistics imply that the structure of the code being produced is significantly better than the code that might have been produced by the programmers working alone.

I haven't seen any studies on mobbing, but the anecdotal evidence[6] is encouraging.

Seniors can collaborate with juniors. When they do, the seniors are slowed down by the juniors for the duration of the session. The juniors, on the other hand, are sped up for the rest of their lives—so it's a good trade-off.

Seniors can collaborate with seniors; just make sure there are no weapons in the room.

Juniors can collaborate with juniors, though seniors should watch such sessions carefully. Juniors are likely to *prefer* to work with other juniors. If that happens too frequently, a senior should step in.

Some folks simply do not like to participate in collaborations like this. Some people work better alone. They should not be forced into collaboration sessions beyond reasonable peer pressure. Nor should they be disparaged for their preference. Often, they will be happier in a mob than in a pair.

Collaboration is a skill that takes time and patience to acquire. Don't expect to be good at it until you've practiced it for many hours. However, it is a skill that is very beneficial to the team as a whole, and to each programmer who engages in it.

6. Agile Alliance, "Mob Programming: A Whole Team Approach," AATC2017, https://www.agilealliance.org/resources/sessions/mob-programming-aatc2017/.

8 ACCEPTANCE TESTS

（验收测试）

Of all the disciplines of clean craftsmanship, acceptance testing is the one that programmers have the least control over. Fulfilling this discipline requires the participation of the business. Unfortunately, many businesses have, so far, proven unwilling to properly engage.

How do you know when a system is ready to deploy? Organizations around the world frequently make this decision by engaging a QA department or group to "bless" the deployment. Typically, this means that the QA folks run a rather large bevy of manual tests that walk through the various behaviors of the system until they are convinced that the system behaves as specified. When those tests "pass," the system may be deployed.

This means that the true requirements of the system *are those tests*. It does not matter what the requirements document says; it is only the tests that matter. If QA signs off after running their tests, the system is deployed. Therefore, it is those tests that are the requirements.

The discipline of acceptance testing recognizes this simple fact and recommends that all requirements be specified *as tests*. Those tests should be written by the business analysis (BA) and QA teams, on a feature-by-feature basis, shortly before each feature is implemented. QA is not responsible for running those tests. Rather, that task is left to the programmers; therefore, the programmers will very likely automate those tests.

No programmer in their right mind wants to manually test the system over and over again. Programmers automate things. Thus, if the programmers are responsible for running the tests, the programmers *will* automate those tests.

However, because BA and QA author the tests, the programmers must be able to prove to BA and QA that the automation actually performs the tests that were authored. Therefore, the language in which the tests are automated must be a language that BA and QA understand. Indeed, BA and QA ought to be able to *write* the tests in that automation language.

Several tools have been invented over the years to help with this problem: FitNesse,[1] JBehave, SpecFlow, Cucumber, and others. But tools are not really the issue. The specification of software behavior is always a simple function of specifying input data, the action to perform, and the expected output data. This is the well-known AAA pattern: Arrange/Act/Assert.[2]

All tests begin by arranging the input data for the test; then the test causes the tested action to be performed. Finally, the test asserts that the output data from that action matches the expectation.

These three elements can be specified in a variety of different ways, but the most easily approachable is a simple tabular format.

widget should render		
wiki text	html text	
normal text	normal text	
this is ''italic'' text	this is <i>italic</i> text	italic widget
this is '''bold''' text	this is <b>bold</b> text	bold widget
!c This is centered text	<center>This is centered text</center>	
!1 header	<h1>header</h1>	
!2 header	<h2>header</h2>	
!3 header	<h3>header</h3>	
!4 header	<h4>header</h4>	
!5 header	<h5>header</h5>	
!6 header	<h6>header</h6>	
http://files/x	<a href="files/x">http://files/x</a>	file link
http://fitnesse.org	<a href="http://fitnesse.org">http://fitnesse.org</a>	http link
SomePage	SomePage<a title="create page" href="SomePage\?edit&nonExistent=true">\[\?\]</a>	missing wiki word

Figure 8.1 A portion of the results of one of the acceptance tests from the FitNesse tool

Figure 8.1, for example, is a portion of one of the acceptance tests within the FitNesse tool. FitNesse is a wiki, and this test checks that the various markup

1. fitnesse.org
2. This pattern is credited to Bill Wake, who identified it in 2001 (https://xp123.com/articles/3a-arrange-act-assert).

gestures are properly translated into HTML. The action to be performed is `widget should render`, the input data is the `wiki text`, and the output is the `html text`.

Another common format is Given-When-Then:

```
Given a page with the wiki text: !1 header
When that page is rendered.
Then the page will contain: <h1>header</h1>
```

It should be clear that these formalisms, whether they are written in an acceptance testing tool or in a simple spreadsheet or text editor, are relatively easy to automate.

The Discipline（纪律）

In the strictest form of the discipline, the acceptance tests are written by BA and QA. BA focuses on the happy path scenarios, whereas QA focuses on exploring the myriad of ways that the system can fail.

These tests are written at the same time as, or just before, the features they test are developed. In an Agile project, divided up into sprints or iterations, the tests are written during the first few days of the sprint. They should all pass by the end of the sprint.

BA and QA provide the programmers with these tests, and the programmers automate them in a manner that keeps BA and QA engaged.

These tests become the *definition of done*. A feature is not complete until all its acceptance tests pass. And when all the acceptance tests pass, the feature is done.

This, of course, puts a huge responsibility on BA and QA. The tests that they write must be full specifications of the features being tested. The suite of acceptance tests *is* the requirements document for the entire system. By writing those tests, BA and QA are certifying that when they pass, the specified features are done and working.

Some BA and QA teams may be unaccustomed to writing such formal and detailed documents. In these cases, the programmers may wish to write the acceptance tests with guidance from BA and QA. The intermediate goal is to create tests that BA and QA *can read* and bless. The ultimate goal is to get BA and QA comfortable enough to write the tests.

THE CONTINUOUS BUILD（持续构建）

Once an acceptance test passes, it goes into the suite of tests that is run during the continuous build.

The continuous build is an automated procedure that runs every time[3] a programmer checks code into the source code control system. This procedure builds the system from source and then runs the suites of automated programmer unit tests and automated acceptance tests. The results of that run are visibly posted, often in an email to every programmer and interested party. The state of the continuous build is something everyone should be continuously aware of.

The continuous running of all these tests ensures that subsequent changes to the system do not break working features. If a previously passing acceptance test fails during the continuous build, the team must immediately respond and repair it before making any other changes. Allowing failures to accumulate in the continuous build is suicidal.

3. Within a few minutes.

II THE STANDARDS

（标准）

Standards are baseline *expectations*. They are the lines in the sand that we decide we cannot cross. They are the parameters we set as the minimum we can accept. Standards can be exceeded, but we must never fail to meet them.

YOUR NEW CTO（你的新 CTO）

Imagine that I am your new CTO. I'm going to tell you what I expect of you. You will read these expectations, and you will view them from two contradictory perspectives.

The first will be the perspective of your managers, executives, and users. And, from their perspective, these expectations will seem obvious and normal. No manager, executive, or user would ever expect less.

As a programmer, you may be more familiar with the second perspective. It is that of the programmers, architects, and technical leads. From their perspective, these expectations will seem extreme, impossible, even insane.

The difference in those two perspectives, the mismatch of those expectations, is the primary failing of the software industry, and it is one that we must urgently repair.

As your new CTO, I expect. . . .

9 PRODUCTIVITY

（生产力）

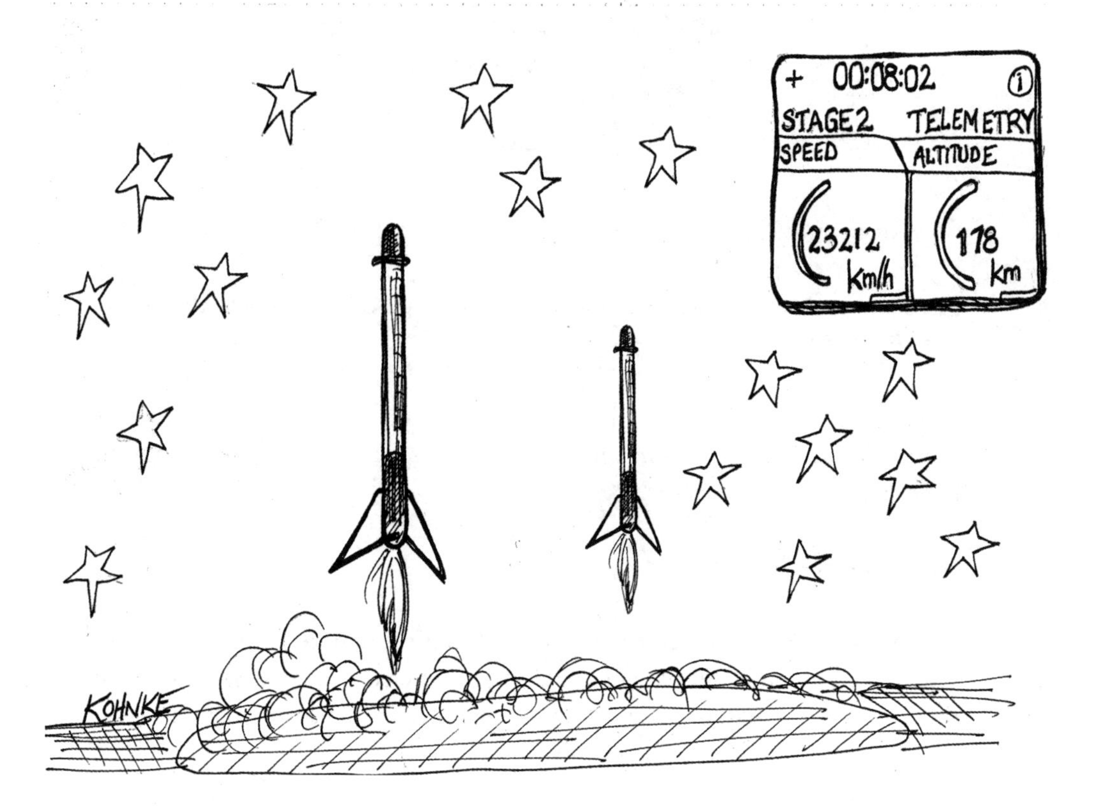

As your CTO, I have a couple of expectations about productivity.

We Will Never Ship S**T（永不交付 S**T）

As your new CTO, I expect that we will never ship S**T.

I'm sure you know what *S**T* stands for. As your CTO, I expect that *we will not ship S**T*.

Have you ever shipped *S**T*? Most of us have. I have. It didn't feel good. I didn't like it. The users didn't like it. The managers didn't like it. *Nobody* liked it.

So why do we do it? Why do we ship *S**T*?

Because somehow or another, we decided that we had no choice. Perhaps it was a deadline that we just absolutely had to meet. Maybe it was an estimate that we were too embarrassed to miss. Maybe it was just downright sloppiness or carelessness. Maybe it was management pressure. Maybe it was a matter of self-worth.

Whatever the reason, it was *invalid*. It is an absolute minimum standard that we will not ship *S**T*.

What is *S**T*? I'm sure you already know, but let's go over it anyway.

- Every bug you ship is *S**T*.
- Every untested function is *S**T*.
- Every poorly written function is *S**T*.
- Every dependency on detail is *S**T*.
- Every unnecessary coupling is *S**T*.
- SQL in the GUI is *S**T*.
- Database schema in the business rules is *S**T*.

I could go on. But let me cut this short. Every failure of one of the disciplines in the foregoing chapters risks shipping *S**T*.

That doesn't mean that every single one of those disciplines must be upheld at all times.

We are engineers. Engineers make trade-offs. But an engineering trade-off is not careless or sloppy. If you must break a discipline, you'd better have a good reason.

More important, you'd better have a good mitigation plan.

For example, let us say that you are writing Cascading Style Sheets (CSS) code. Writing automated tests up front for CSS is almost always impractical. You don't really know how the CSS will render until you actually see it on the screen.

So how are we going to mitigate the fact that CSS breaks the test-driven development testing discipline?

We're going to have to test the CSS manually, with our eyeballs. We're also going to have to test it in all the browsers that our customers are likely to use. So, we'd better come up with a standard description of what we want to see on the screen and how much variation we can tolerate. Even more important, we'd better come up with a technical solution that makes that CSS *easy* to manually test because *we*, not QA, are going to test it before we ever release it from development.

Let me say this all another way: *Do a good job!*

That's what everyone really expects. All our managers, all our users, everyone who ever touches or is touched by our software, expects us to have done a good job. We must not let them down.

I expect we will never ship S**T.

INEXPENSIVE ADAPTABILITY（成本低廉的变更适应能力）

Software is a compound word that means "flexible product." The entire reason that software exists is so that we can quickly and easily change the behavior of our machines. To the extent that we build software that is hard to change, we thwart the very reason that software exists.

And yet inflexibility of software remains a huge problem in our industry. The reason we focus so much on design and architecture is to improve the flexibility and maintainability of our systems.

Why does software become rigid, inflexible, and fragile? Again, it is because software teams fail to engage in the testing and refactoring disciplines that support flexibility and maintainability. In some cases, those teams may depend solely on initial design and architecture efforts. In other cases, they may depend on fads that make unsustainable promises.

But no matter how many microservices you create, no matter how well structured your initial design and architectural vision is, without the testing and refactoring disciplines, the code will rapidly degrade and the system will become harder and harder to maintain.

I do not expect this. I expect that when customers ask for changes, the development team will be able to respond with a strategy that involves an expense that is *proportional to the scope of the change.*

Customers may not understand the internals of the system, but they have a good sense of the scope of the changes they request. They understand that a change may affect many features. They expect the cost of that change to be relative to the scope of that change.

Unfortunately, too many systems become so inflexible over time that the cost of change rises to a level that customers and managers cannot rationalize against the scope of the changes they request. To make matters worse, it is not uncommon for developers to rail against certain kinds of change on the basis that the change is against the architecture of the system.

An architecture that resists the changes that a customer requests is an architecture that thwarts the meaning and intent of software. Such an architecture must be changed to accommodate the changes that the customers will make. Nothing makes such changes easier than a well-refactored system and a suite of tests that you trust.

I expect that the design and architecture of the system will evolve with the requirements. I expect that when customers request changes, those changes will not be impeded by existing architecture or the rigidity and fragility of the existing system.

I expect inexpensive adaptability.

We Will Always Be Ready（时刻准备着）

As your new CTO, I expect that we will always be ready.

Long before Agile made it popular, it was well understood by most software experts that well-run projects experience a regular rhythm of deployment and release. In the very early days, this rhythm tended to be quick: weekly or even daily. However, the waterfall movement that began in the 1970s greatly slowed the rhythm into months, sometimes years.

The advent of Agile, at the turn of the millennium, reasserted the need for faster rhythms. Scrum recommended sprints of 30 days. XP recommended iterations of 3 weeks. Both quickly increased the rate to biweekly. Nowadays, it is not uncommon for development teams to deploy multiple times per day, effectively reducing the development period to near zero.

I expect a quick rhythm. One or two weeks at most. And at the end of each sprint, I expect that the software will be technically ready to release.

Technically ready to release does not mean that the business will *want* to release it. The technically ready software may not have a feature set that the business deems complete or appropriate for its customers and users. Technically ready simply means that if the business decides to release it, the development team, including QA, has no objections. The software works, has been tested, has been documented, and is ready for deployment.

This is what it means to *always be ready*. I do not expect the development team to tell the business to wait. I do not expect long burn-in periods or so-called *stabilization sprints*. Alpha and beta testing may be appropriate to determine feature compatibility with users but should not be used to drive out coding defects.

Long ago, my company consulted for a team who built word processors for the legal profession. We taught them Extreme Programming. They eventually got to the point that every week the team would burn a new CD.[1] They'd put that CD on top of a stack of weekly releases that was kept in the developer's lab. The salespeople, on their way to do a demo for a prospective customer, would walk into the lab and take the top CD on the stack. That's how *ready* the development team was. That's how ready I expect us to be.

Being ready as frequently as this requires a very high discipline of planning, testing, communication, and time management. These are, of course, the disciplines of Agile. Stakeholders and developers must be frequently engaged in estimating and selecting the highest-value development stories. QA must be deeply engaged in providing automated acceptance tests that define "done." Developers must work closely together and maintain an intense testing, review, and refactoring discipline in order to make progress in the short development periods.

But *always being ready* is more than just following the dogma and rituals of Agile. Always being ready is an attitude, a way of life. It is a commitment to continuously provide incremental value.

I expect that we will always be ready.

Stable Productivity（稳定的生产力）

Software projects often experience a decrease in productivity with time. This is a symptom of serious dysfunction. It is caused by the neglect of testing and refactoring disciplines. That neglect leads to the ever-increasing impediment of tangled, fragile, and rigid code.

This impediment is a runaway effect. The more fragile and rigid the code within a system becomes, the more difficult that code is to keep clean. As the fragility of the code increases, so does the fear of change. Developers become ever-more reluctant to clean up messy code because they fear that any such effort will lead to more defects.

1. Yes, there was a time in the deep dark past when software was distributed on CDs.

This process leads, in a matter of months, to an extreme and accelerating loss of productivity. Each succeeding month sees the team's productivity seeming to approach zero asymptotically.

Managers often try to combat this decline in productivity by adding manpower to the project. But this strategy often fails because the new programmers brought onto the team are no less subject to the fear of change than the programmers who have been there all along. They quickly learn to emulate the behavior of those team members and thereby perpetuate the problem.

When pressed about the loss of productivity, the developers often complain about the awful nature of the code. They may even begin to militate for a redesign of the system. Once begun, this complaint grows in volume until managers cannot ignore it.

The argument posed by the developers is that they can increase productivity if they redesign the system from scratch. They argue that they know the mistakes that were made and will not repeat them. Managers, of course, do not trust this argument. But managers are desperate for anything that will increase productivity. In the end, many managers accede to the demands of the programmers despite the costs and risks.

I do not expect this to happen. I expect development teams to keep their productivity consistently high. I expect that development teams will reliably employ the disciplines that keep the structure of the software from degrading.

I expect stable productivity.

10 QUALITY

（质量）

As your CTO, I have several expectations about quality.

Continuous Improvement（持续改进）

I expect continuous improvement.

Human beings improve things with time. Human beings impose order upon chaos. Human beings make thing better.

Our computers are better than they used to be. Our cars are better than they used to be. Our airplanes, our roads, our telephones, our TV service, our communications services are all better than they used to be. Our medical technology is better than it used to be. Our space technology is better than it used to be. Our civilization is massively better than it used to be.

Why, then, does software degrade with time? I do not expect our software to degrade.

I expect that, as time goes by, the design and architecture of our systems will improve. I expect that the software will get cleaner and more flexible with every passing week. I expect that the cost of change will *decrease* as the software ages. I expect everything to get better with time.

What does it take to make software better with time? It takes will. It takes attitude. It takes a commitment to the disciplines that we *know* work.

I expect that every time any programmer checks in code, they check it in cleaner than they checked it out. I expect that every programmer *improves* the code that they touch, regardless of why they are touching it. If they fix a bug, they should also make the code better. If they are adding a feature, they should also make the code better. I expect every manipulation of the code to result in better code, better designs, and better architecture.

I expect continuous improvement.

FEARLESS COMPETENCE（免于恐惧）

I expect fearless competence.

As the internal structure of a system degrades, the complexity of the system can rapidly become intractable. This causes the developers to naturally become more and more fearful of making changes. Even simple improvements become fraught with risk. Reluctance to make changes and improvements can drastically lower the programmers' competence to manage and maintain the system.

This loss of competence is not intrinsic. The programmers have not become less competent. Rather, the growing intractable complexity of the system begins to exceed the natural competence of the programmers.

As the system becomes increasingly difficult for the programmers to handle, they begin to fear working on it. That fear exacerbates the problem because programmers who are afraid to change the system will only make changes that they feel are safest to make. Such changes are seldom those that improve the system. Indeed, often the so-called safest changes are those that degrade the system even more.

If this trepidation and timidity are allowed to continue, estimates will naturally grow, defect rates will increase, deadlines will become harder and harder to achieve, productivity will plumet, and morale will slide into the pit.

The solution is to eliminate the fear that accelerates the degradation. We eliminate the fear by employing the disciplines that create suites of tests that the programmers trust with their lives.

With such tests in place and with the skill to refactor and drive toward simple design, the programmers will not fear to clean a system that has degraded in one place or another. They will have the confidence and the competence to quickly repair that degradation and keep the software on an ever-improving trajectory.

I expect the team to always exhibit fearless competence.

Extreme Quality（极致质量）

I expect extreme quality.

When did we first begin to accept that bugs are just a natural part of software? When did it become acceptable to ship software with a certain level of defects? When did we decide that beta tests were appropriate for general distribution?

I do not accept that bugs are unavoidable. I do not accept the attitude that expects defects. I expect every programmer to deliver software that is *defect free.*

And I am not simply referring to behavior defects. I expect every programmer to deliver software that is free of defects in behavior *and structure.*

Is this an achievable goal? Can this expectation be met? Whether it can or can't, I expect every programmer to accept it as the standard and continuously work toward achieving it.

I expect extreme quality coming out of the programming team.

We Will Not Dump on QA（我们不把问题留给 QA）

I expect that we will not dump on QA.

Why do QA departments exist? Why would companies invest in entirely separate groups of people to check the work of the programmers? The answer is obvious and depressing. Companies chose to create software QA departments because programmers were not doing their jobs.

When did we get the idea that QA belongs at the end of the process? In too many organizations, QA sits waiting for the programmers to release the software to them. Of course, the programmers do not release the software on schedule, so it is QA that is stuck trying to make the release date by cutting their testing short.

This puts the QA people under tremendous pressure. It is a high-stress, tedious job in which shortcuts are necessary if shipment dates are to be maintained. And that is clearly no way to assure quality.

The QA Disease（QA之疾）

How do you know if QA is doing a good job? On what basis do you give them raises and promotions? Is it defect discovery? Are the best QA people the ones who find the most defects?

If so, then QA views defects as positive things. The more the better! And that is, of course, sick.

But QA may not be the only folks who view defects in a positive light. There's an old saying[1] in software: "I can meet any schedule you like, so long as the software doesn't need to work."

That may sound funny, but it is also a strategy that developers can use to meet individual deadlines. If it's QA's job to find the bugs, why not deliver on time and let them find some?

No word has to be spoken. No deal has to be made. No handshakes ever occur. And yet everyone knows that there is an economy of bugs flowing back and forth between the developers and QA. And that is a very deep sickness.

I expect that we will not dump on QA.

QA Will Find Nothing（QA什么问题也不会发现）

I expect that if QA is at the end of the process, then QA will find nothing. It should be the goal of the development team that QA at the end never finds a bug. Anytime a bug is found by QA, the developers should determine to find out why, correct the process, and make sure it never happens again.

1. I first heard it from Kent Beck.

QA should wonder why they are at the end of the process, because they never find anything.

In fact, QA does not belong at the end of the process. QA belongs at the beginning of the process. The job of QA is not to find all the bugs; that's the programmers' job. The job of QA is to specify the system behavior in terms of tests with sufficient detail that defects are excluded from the final system. Those tests should be executed by the programmers, not QA.

I expect that QA will find nothing.

Test Automation（测试自动化）

In most instances, manual testing is a huge waste of money and time. Almost any test that can be automated *should* be automated. This includes unit tests, acceptance tests, integration tests, and system tests.

Manual testing is expensive. It should be reserved for situations in which human judgment is necessary. This includes things like checking the aesthetics of a GUI, exploratory testing, and subjectively evaluating the ease of an interaction.

Exploratory testing deserves a special mention. This kind of testing depends entirely on human ingenuity, intuition, and insight. The goal is to empirically derive the behavior of the system through extensive observation of the way the system operates. Exploratory testers must infer corner cases and deduce appropriate operational pathways to exercise them. This is no mean feat and requires a significant amount of expertise.

Most tests, on the other hand, are eminently automatable. The vast majority are simple Arrange/Act/Assert constructs that can be executed by supplying canned inputs and examining expected outputs. Developers are responsible to provide a function-callable API that allows these tests to run quickly and without significant setup of an execution environment.

Developers should design the system to abstract out any slow or high-setup operations. If, for example, the system makes extensive use of a relational database management system (RDBMS), the developers should create an abstraction layer that encapsulates the business rules from it. This practice allows the automated tests to replace the RDBMS with canned input data, vastly increasing both the speed and reliability of the tests.

Slow and inconvenient peripherals, interfaces, and frameworks should also be abstracted so that individual tests can run in microseconds, can be run in isolation from any environment,[2] and are not subject to any ambiguity regarding socket timing, database contents, or framework behavior.

Automated Testing and User Interfaces (自动化测试与用户界面)

Automated tests *should not* test business rules through the user interface. User interfaces are subject to change for reasons that have more to do with fashion, facility, and general marketing chaos than with business rules. When automated tests are driven through the user interface, as shown Figure 10.1, those tests are subject to those changes. As a consequence, the tests become very fragile, which often results in the tests being discarded as too difficult to maintain.

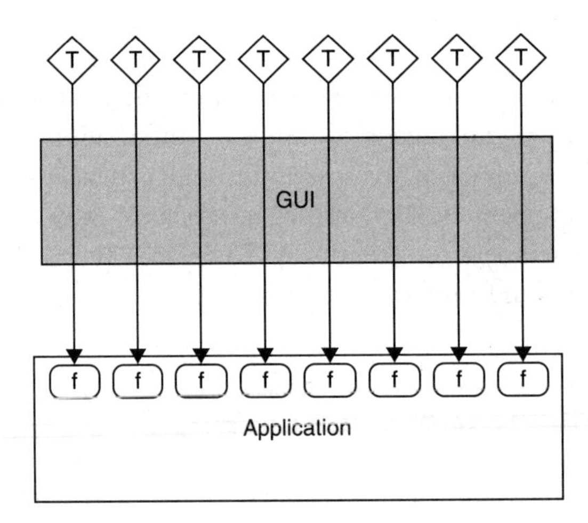

Figure 10.1 Tests driven through the user interface

2. For example, at 30,000 feet over the Atlantic on your laptop.

To avoid this situation, developers should isolate the business rules from the user interface with a function call API, as shown in Figure 10.2. Tests that use this API are completely independent of the user interface and are not subject to interface changes.

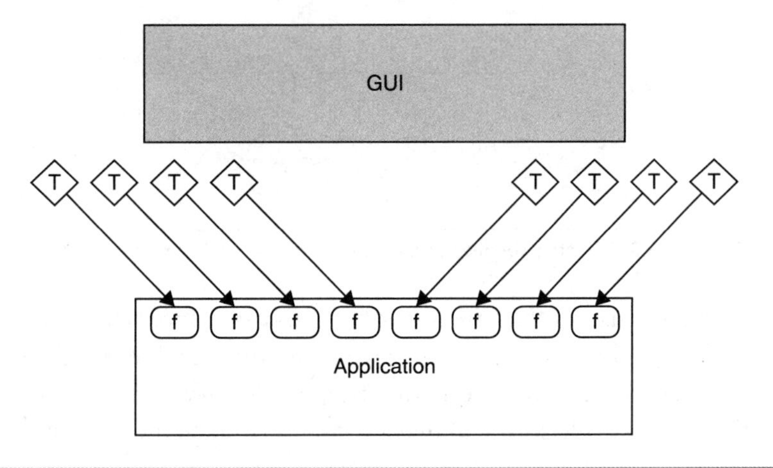

Figure 10.2 Tests through the API are independent of the user interface.

Testing the User Interface（测试用户界面）

If the business rules are automatically tested through a function call API, then the amount of testing required for the behavior of the user interface is vastly reduced. Care should be taken to maintain the isolation from the business rules by replacing the business rules with a stub that supplies canned values to the user interface, as shown in Figure 10.3.

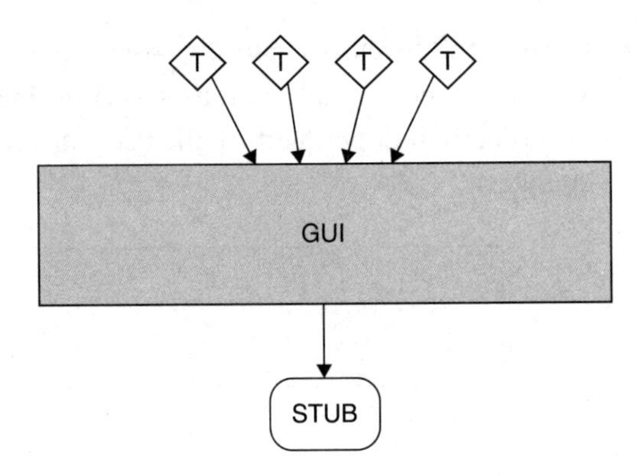

Figure 10.3 The stub supplies canned values to the user interface.

Doing so ensures that the tests of the user interface will be fast and unambiguous. If the user interface is significantly large and complex, then an automated user interface testing framework may be appropriate. The use of the business rule stub will make those tests much more reliable.

If the user interface is small and simple, it may be more expeditious to fall back on manual testing, especially if there are aesthetics to be evaluated. Again, the use of the business rule stub will make these manual tests much easier to conduct.

I expect that every test that can practicably be automated *will* be automated, that the tests will execute quickly, and that the tests will not be fragile.

11 COURAGE

（勇气）

As CTO, I have several expectations related to courage.

We Cover for Each Other（我们彼此补位）

We use the word *team* to describe a group of developers working on a project. But do we understand what a team really is?

A team is a group of collaborators who understand their goals and their interaction so well that when a team member goes down for some reason, *they keep making progress toward their goal*. For example, every crew member on board a ship has a job to do. Every crew member also knows how to do someone else's job—for obvious reasons. The ship has to keep sailing even when a crew member goes down.

I expect that the members of a programming team will cover for each other like the crew of a ship. When a team member goes down, I expect others on the team to take over that role until the fallen team member resumes their place on the team.

People on a team can go down for many reasons. They may get sick. They may be distracted by trouble at home. They may go on vacation. Work on the project cannot stop. Others must fill the hole left behind.

If Bob is the database guy, and Bob goes down, someone else must pick up the database work and keep making progress. If Jim is the GUI guy, and Jim goes down, someone else must pick up the GUI work and continue to make progress.

This means that each member of the team must be familiar with more than just their own work. They must be familiar with the work of others so that they can step in if one of those others goes down.

But let me turn this around. It is *your* responsibility to make sure someone can cover for *you*. It is your responsibility to ensure that you are not the one indispensable player on the team. It is your responsibility to seek out others and teach them enough about your work that they can take over for you in a pinch.

How can you teach others about your work? Probably the best way is to sit down with them at a workstation and write code together for an hour or so. And if you are wise, you will do this with more than one other member of the team. The more people who know your work, the more people can cover for you if you go down.

And remember, once is not enough. As you continue to make progress on your part of the project, you'll have to continually keep the others abreast of your work.

You will find the discipline of collaborative programming helpful in this regard.

I expect that the members of programming teams will be able to cover for each other.

Honest Estimates（靠谱的预估）

I expect honest estimates.

As a programmer, the most honest estimate you can give is “I don’t know” because you actually do not know how long the task will take. On the other hand, you *do* know that you will probably finish the task in less than a billion years. So, an honest estimate is an amalgam of what you don’t know with what you do know.

An honest estimate looks something like this:

- I have a 5 percent chance of finishing this task before Friday.
- I have a 50 percent chance of finishing before the next Friday.
- I have a 95 percent chance of finishing before the Friday after that.

An estimate like this provides a probability distribution that describes your uncertainty. Describing your uncertainty is what makes this estimate honest.

You should provide estimates in this form when managers ask you to estimate large projects. For example, they may be trying to judge the cost of a project before they authorize it. That's when this kind of honesty about uncertainty is most valuable.

For smaller tasks, it is best to use the Agile practice of story points. Story points are honest because they do not commit to a timeframe. Rather, they describe the cost of a task in comparison to another. The numbers used are arbitrary, but relative.

A story point estimate looks something like this:

> *The Deposit story has a cost of 5.*

What is that 5? It's an arbitrary number of points that is relative to some task of known size. For example, let's say that the Login story was arbitrarily given 3 points. When you estimate the Deposit story, you decide that Deposit is not quite twice as hard as Login, so you give it a 5. That's really all there is to it.

Story points already have the probability distribution embedded within them. First, the points are not dates or times; they are just points. Second, the points are not promises; they are guesses. At the end of each Agile iteration (usually a week or two), we total the points completed. We use that number to estimate how many points we might complete in the next iteration.

I expect honest estimates that describe your uncertainty. I do not expect a promise of a date.

You Must Say NO（你得说不）

I expect you to say no when the answer is no.

One of the most important things a programmer can say is "No!" Said at the right time, in the right context, this answer can save your employer massive amounts of money and prevent horrible failures and embarrassments.

This is not a license to storm around saying no to everything. We are engineers; our job is to find a way to yes. But sometimes *yes* is not an option. We are the only ones who can determine this. We are the ones who know. Therefore, it is up to us to say no when the answer really is no.

Let's say your boss asks you to get something done by Friday. After giving it due consideration, you realize that there is no reasonable chance that you'll complete the task by Friday. You must return to your boss and say "No." You would be wise to also say that you can get it done by the following Tuesday, but you must be firm that Friday is out of the question.

Managers often don't like to hear no. They may push back on you. They may confront you. They may yell at you. Emotional confrontation is one of the tools that some managers employ.

You must not give in to it. If the answer is no, then you must hold to that answer and not yield to the pressure.

And be very wary of the "Will you at least try?" gambit. It seems so reasonable to be asked to try, doesn't it? But it's not reasonable at all because *you are already trying*. There's nothing new you can do to change the no to yes, so saying you'll try is just a lie.

I expect that when the answer is no, you will say no.

Continuous Aggressive Learning（持续努力学习）

The software industry is wildly dynamic. We can debate whether this *should* be so, but we cannot debate whether it *is* so. It is. And therefore, we must all be continuous aggressive learners.

The language you are using today will likely not be the language you'll be using in 5 years. The framework you are using today will probably not be the framework you'll be using next year. Be prepared for these changes by being aware of what is changing all around you.

Programmers have often been advised[1] to learn a new language every year. This is good advice. Moreover, pick a language that has a style you are unfamiliar with. If you've never written code in a dynamically typed language, learn one. If you've never written code in a declarative language, learn one. If you've never written Lisp or Prolog or Forth, learn them.

How and when do you do this learning? If your employer provides you the time and space to do this kind of learning, then take as much advantage of it

1. David Thomas and Andrew Hunt, *The Pragmatic Programmer: From Journey to Mastery* (Addison-Wesley, 2020).

as you can. If your employer is not so helpful, then you'll have to learn on your own time. Be prepared to spend several hours per month on it. Make sure you have the personal time set aside for it.

Yes, I know, you have family obligations, there are bills to pay and planes to catch, and you've got a *life*. Okay, but you also have a *profession*. And professions need care and maintenance.

I expect us all to be continuous aggressive learners.

MENTORING（教导）

We seem to have an unending need for more and more programmers. The number of programmers in the world is increasing at a furious and exponential pace. Universities can only teach so much, and unfortunately, many of them fail to teach much at all.

Therefore, the job of teaching new programmers falls to us. We, the programmers who have been working for a few years, must pick up the burden of teaching those who have just started.

Perhaps you think this is hard. It is. But it comes with a huge benefit. The best way to learn is to teach. Nothing else even comes close. So, if you want to learn something, teach it.

If you have been a programmer for 5 years, or 10 years, or 15 years, you have an immense amount of experience and life lessons to teach to the new programmers who have just started. Take one or two of them under your wing and guide them through their first 6 months.

Sit down with them at their workstations and help them write code. Tell them stories about your past failures and successes. Teach them about disciplines, standards, and ethics. Teach them the craft.

I expect all programmers to become mentors. I expect you to get involved in helping others to learn.

III

THE ETHICS

（操守）

The First Programmer（第一个程序员）

The profession of software began, inauspiciously, in the summer of 1935 when Alan Turing began work on his paper. His goal was to resolve a mathematical dilemma that had perplexed mathematicians for a decade or more—the Entscheidungsproblem. The decision problem.

In that goal he was successful, but he had no idea, at the time, that his paper would spawn a globe-spanning industry upon which we would all depend and that now forms the life's blood of our entire civilization.

Many people think of Lord Byron's daughter, Ada, the countess of Lovelace, as the first programmer, and with good reason. She was the first person we know of who understood that the numbers manipulated by a computing machine could represent nonnumeric concepts. Symbols instead of numbers. And, to be fair, Ada did write some algorithms for Charles Babbage's Analytical Engine, which unfortunately was never built.

But it was Alan Turing who wrote the first[1] programs to execute in an electronic computer. And it was Alan Turing who first *defined* the profession of software.

In 1945, Turing wrote code for the Automated Computing Engine (ACE). He wrote this code in binary machine language, using base-32 numbers. Code like this had never been written before, so he had to invent and implement concepts such as subroutines, stacks, and floating-point numbers.

After several months of inventing the basics and using them to solve mathematical problems, he wrote a report that stated the following conclusion:

> *We shall need a great number of mathematicians of ability, because there will probably be a good deal of work of this kind to be done.*

---

1. Some folks point out that Konrad Zuse wrote algorithms for his electromechanical computer before Turing programmed the ACE.

"A great number." How did he know? In reality, he had no idea just how prescient that statement was—we certainly have a great number now.

But what was that other thing he said? "Mathematicians of ability." Do you consider yourself to be a mathematician of ability?

In the same report, he went on to write:

> *One of our difficulties will be the maintenance of an appropriate discipline, so that we do not lose track of what we are doing.*

"Discipline!" How did he know? How could he look forward 70 years and know that our problem would be discipline?

Seventy years ago, Alan Turing laid the first stone in the framework of software professionalism. He said that we should be mathematicians of ability who maintain an appropriate discipline.

Is that who we are? Is that who you are?

Seventy-Five Years (75年)

One person's lifetime. That's how old our profession is as of this writing. Just 75 years. And what has happened in those three score and fifteen years? Let's more carefully revisit the history that I presented in Chapter 1, "Craftsmanship."

In 1945, there was one computer in the world and one programmer. Alan Turing. These numbers grew rapidly in those first years. But let's use this as our origin.

> *1945. Computers: O(1). Programmers: O(1).*

In the decade that followed, the reliability, consistency, and power usage of vacuum tubes improved dramatically. This made it possible for larger and more powerful computers to be built.

By 1960, IBM had sold 140 of its 700 series computers. These were huge, expensive behemoths that could only be afforded by military, government, and very large corporations. They were also slow, resource limited, and fragile.

It was during this period that Grace Hopper invented the concept of a higher-level language and coined the term *compiler*. By 1960, her work led to COBOL.

In 1952, John Backus submitted the FORTRAN specification. This was followed rapidly by the development of ALGOL. By 1958, John McCarthy had developed LISP. The proliferation of the language zoo had begun.

In those days, there were no operating systems, no frameworks, no subroutine libraries. If something executed on your computer, it was because you wrote it. Consequently, in those days, it took a staff of a dozen or more programmers just to keep one computer running.

By 1960, 15 years after Turing, there were O(100) computers in the world. The number of programmers was an order of magnitude greater: O(1,000).

Who were these programmers? They were people like Grace Hopper, Edsger Dijkstra, Jon Von Neumann, John Backus, and Jean Jennings. They were scientists and mathematicians and engineers. Most were people who already had careers and already understood the businesses and disciplines they were employed by. Many, if not most, were in their 30s, 40s, and 50s.

The 1960s was the decade of the transistor. Bit by bit, these small, simple, inexpensive, and reliable devices replaced the vacuum tube. And the effect on computers was a game changer.

By 1965, IBM had produced more than ten thousand 1401 transistor-based computers. They rented for about $2,500 per month, putting them within reach of thousands of medium-sized businesses.

These machines were programmed in Assembler, Fortran, COBOL, and RPG. And all those companies who rented those machines needed staffs of programmers to write their applications.

IBM wasn't the only company making computers at the time, so we'll just say that by 1965, there were O(10,000) computers in the world. And if each computer needed ten programmers to keep it running, there must have been O(100,000) programmers.

Twenty years after Turing, there must have been several hundred thousand programmers in the world. Where did these programmers come from? There weren't enough mathematicians, scientists, and engineers to cover the need. And there weren't any computer science graduates coming out of the universities because there weren't any computer science degree programs—anywhere.

Companies thus drew from the best and brightest of their accountants, clerks, planners, and so on—anyone with some proven technical aptitude. And they found lots.

And, again, these were people who were already professionals in another field. They were in their 30s and 40s. They already understood deadlines, and commitments, what to leave in, and what to leave out.[2] Although these people were not mathematicians per se, they were disciplined professionals. Turing would likely have approved.

But the crank kept on turning. By 1966, IBM was producing a thousand 360s every month. These computers were popping up everywhere. They were immensely powerful for the day. The model 30 could address 64K bytes of memory and execute 35,000 instructions per second.

It was during this period, in the mid-1960s, when Ole Johann Dahl and Kristen Nygard invented Simula-67—the first object-oriented language. It was also during this period that Edsger Dijkstra invented structured programming.

2. Apologies to Bob Seger.

And it was also during this time that Ken Thompson and Dennis Ritchie invented C and UNIX.

Still the crank turned. In the early 1970s, the integrated circuit came into regular use. These little chips could hold dozens, hundreds, even thousands of transistors. They allowed electronic circuits to be massively miniaturized.

And thus, the minicomputer was born.

In the late 1960s and into the 1970s, Digital Equipment Corporation sold fifty thousand PDP-8 systems and hundreds of thousands of PDP-11 systems.

And they weren't alone! The minicomputer market exploded. By the mid-1970s, there were dozens and dozens of companies selling minicomputers, so by 1975, 30 years after Turing, there were about 1 million computers in the world. And how many programmers were there? The ratio was starting to change. The number of computers was approaching the number of programmers, so by 1975, there were O(1E6) programmers.

Where did these millions of programmers come from? Who were they?

They were me. Me and my buddies. Me and my cohort of young, energetic, geeky boys.

Tens of thousands of new electronic engineering and computer science grads: We were all young. We were all smart. We, in the United States, were all concerned about the draft. And we were almost all male.

Oh, it's not that women were leaving the field in any number—yet. That didn't start until the mid-1980s. No, it's just that many more boys (and we were boys) were entering the field.

In my first job as a programmer, in 1969, there were a couple of dozen programmers. They were all in their 30s or 40s, and a third to a half were women.

Ten years later, I was working at a company with about 50 programmers, and perhaps three were women.

So, 30 years after Turing, the demographics of programming had shifted dramatically toward very young men. Hundreds of thousands of twenty-something males. We were typically *not* what Turing would have described as disciplined mathematicians.

But businesses had to have programmers. The demand was through the roof. And what very young men lack in discipline, they make up for with energy.

We were also cheap. Despite the high starting salaries of programmers today, back then companies could pick up programmers pretty inexpensively. My starting salary in 1969 was $7,200 per year.

This has been the trend ever since. Young men have been pouring out of computer science programs every year, and industry seems to have an insatiable appetite for them.

In the 30 years between 1945 and 1975, the number of programmers grew by at least a factor of a million. In the 40 years since then, that growth rate has slowed a bit but is still very high.

How many programmers do you think were in the world by 2020? If you include the VBA[3] programmers, I think there must be hundreds of millions of programmers in the world today.

This is clearly exponential growth. Exponential growth curves have a doubling rate. You can do the math. Hey, Albert, what's the doubling rate if we go from 1 to 100 million in 75 years?

The log to the base 2 of 100 million is approximately 27—divide that into 75, and you get about 2.8, so perhaps the number of programmers doubled roughly every two and a half-ish years.

---

3. Visual Basic for Applications.

Actually, as we saw earlier, the rate was higher in the first decades and has slowed down a bit now. My guess is that the doubling rate is about five years. Every five years, the number of programmers in the world doubles.

The implications of that fact are staggering. If the number of programmers in the world doubles every five years, it means that half the programmers in the world have less than five years' experience, and this will always be true so long as that doubling rate continues. This leaves the programming industry in the precarious position of—perpetual inexperience.

Nerds and Saviors（书呆子与救世主）

Perpetual inexperience. Oh, don't worry, this doesn't mean that *you* are perpetually inexperienced. It just means that once you gain 5 years of experience, the number of programmers will have doubled. By the time you gain 10 years of experience, the number of programmers will have quadrupled.

People look at the number of young people in programming and conclude that it's a young person's profession. They ask, "Where are all the old people?"

We're all still here! We haven't gone anywhere. There just weren't that many of us to begin with.

The problem is that there aren't enough of us old guys to teach the new programmers coming in. For every programmer with 30 years' experience, there are 63 programmers who need to learn something from her (or him), 32 of whom are brand new.

Hence the state of perpetual inexperience, with insufficient mentors to correct the problem. The same old mistakes get repeated over and over and over again.

But something else has happened in the last 70 years. Programmers have gained something that I am sure Alan Turing never anticipated: notoriety.

Back in the 1950s and 1960s, nobody knew what a programmer was. There weren't enough of them to have a social impact. Programmers did not live next door to very many people.

That started to change in the 1970s. By then, fathers were advising their sons (and sometimes their daughters) to get degrees in computer science. There were enough programmers in the world so that everybody knew somebody who knew one. And the image of the nerdy, twinkie-eating geek was born.

Few people had seen a computer, but virtually everyone had heard about them. Computers showed up in TV shows such as *Star Trek* and movies such as *2001: A Space Odyssey* and *Colossus: The Forbin Project*. All too often in those shows, the computers were cast as villains. But in Robert Heinlein's 1966 book *The Moon Is a Harsh Mistress*,[4] the computer was the self-sacrificing hero.

Notice, however, that in each of these cases, the programmer is not a significant character. Society didn't know what to make of programmers back then. They were shadowy, hidden, and somehow insignificant compared to the machines themselves.

I have fond memories of one television commercial from this era. A wife and her husband, a nerdy little guy with glasses, a pocket protector, and a calculator, were comparing prices at a grocery store. Mrs. Olsen described him as "a computer genius" and proceeded to school the wife and husband alike on the benefits of a particular brand of coffee.

The computer programmer in that commercial was naive, bookish, and inconsequential. Someone smart, but with no wisdom or common sense. Not someone you'd invite to parties. Indeed, computer programmers were seen as the kind of people who got beaten up a lot at school.

---

4. Robert Heinlein, *The Moon Is a Harsh Mistress* (Ace, 1966).

By 1983, personal computers started to appear, and it was clear that teenagers were interested in them for lots of reasons. By this time, a rather large number of people knew at least one computer programmer. We were considered professionals but still mysterious.

That year, the movie *War Games* depicted a young Mathew Broderick as a computer-savvy teenager and hacker. He hacks into the United States' weapons control system, thinking it is a video game, and starts the countdown to thermonuclear war. At the end of the movie, he saves the world by convincing the computer that the only winning move is not to play.

The computer and the programmer had switched roles. Now it was the computer that was the childlike naive character and the programmer the conduit, if not the source, of wisdom.

We saw something similar in the 1986 movie *Short Circuit* in which the computerized robot known as Number 5 is childlike and innocent but learns wisdom with the help of its creator/programmer and his girlfriend.

By 1993, things had changed dramatically. In the film *Jurassic Park,* the programmer was the villain, and the computer was not a character at all. It was just a tool.

Society was beginning to understand who we were and the role we played. We had graduated from nerd to teacher to villain in just 20 years.

But the vision changed again. In the 1999 film *The Matrix,* the main characters were both programmers and saviors. Indeed, their godlike powers came from their ability to read and understand—"the code."

Our roles were changing fast. Villain to savior in just a few years. Society at large was beginning to understand the power we have both for good and for bad.

Role Models and Villains（榜样和恶棍）

Fifteen years later, in 2014, I visited the Mojang office in Stockholm to do some lectures on clean code and test-driven development. Mojang, in case you didn't know, is the company that produced the game Minecraft.

Afterward, because the weather was nice, the Mojang programmers and I sat outside at a beer garden, chatting. All of a sudden, a young boy, perhaps 12, runs up to the fence and calls out to one of the programmers: "Are you Jeb?"

He was referring to Jens Bergensten, one of the lead programmers at Mojang.

The lad asked Jens for his autograph and peppered him with questions. He had eyes for no one else.

And I'm, like, sitting right there. . . .

Anyway, the point is that programmers have become role models and idols for our children. They dream of growing up to be like Jeb or Dinnerbone or Notch.

Programmers, real-life programmers, are heroes.

But where there are real heroes, there are also real villains.

In October of 2015, Michael Horn, the CEO of Volkswagen North America, testified before the US Congress regarding the software in their cars that was cheating the Environmental Protection Agency's testing devices. When asked why the company did this, he blamed programmers. He said, "This was a couple of software engineers who put this in for whatever reasons."

Of course, he was lying about the "whatever reasons." He knew what the reasons were, and so did the Volkswagen company at large. His feeble attempt to shift blame onto the programmers was pretty transparent.

On the other hand, he was exactly right. It *was* some programmers who wrote that lying, cheating code.

And those programmers—whoever they were—gave us all a bad name. If we had a true professional organization, their recognition as programmers would, and should, be revoked. They betrayed us all. They besmirched the honor of our profession.

And so, we've graduated. It's taken 75 years. But we've gone from nothing to nerds to role models and villains in that time.

Society has begun—just begun—to understand who we are and the threats and promises that we represent.

We Rule the World（我们统治世界）

But society doesn't understand everything yet. Indeed, neither do we. You see, you and I, we are programmers, and we rule the world.

That may seem like an exaggerated statement, but consider. There are more computers in the world, right now, than there are people. And these computers, which outnumber us, perform myriads of essential tasks for us. They keep track of our reminders. They manage our calendars. They deliver our Facebook messages and keep our photo albums. They connect our phone calls and deliver our text messages. They control the engines in our cars, as well as the brakes, accelerator, and sometimes even the steering wheels.

We can't cook without them. We can't wash clothes without them. They keep our houses warm in winter. They entertain us when we're bored. They keep track of our bank records and our credit cards. They help us pay our bills.

In fact, most people in the modern world interact with some software system every waking minute of every day. Some even continue interacting while they sleep.

The point is that *nothing* happens in our society without software. No product gets bought or sold. No law gets enacted or enforced. No car drives. No Amazon products get delivered. No phone connects. No power comes out of outlets. No food gets delivered to stores. No water comes out of faucets. No gas gets piped to furnaces. None of these things happens without software monitoring and coordinating it all.

And *we* write that software. And that makes us the rulers of the world.

Oh, other people think they make the rules—but then they hand those rules to us, and *we* write the rules that execute in the machines that monitor and coordinate every aspect of our lives.

Society does not quite understand this yet. Not quite. Not yet. But the day is coming soon when our society will understand it all too well.

We, programmers, don't quite understand this yet either. Not really. But, again, the day is coming when it will be savagely driven home to us.

Catastrophes（灾难）

We've seen plenty of software catastrophes over the years. Some have been pretty spectacular.

For example, in 2016, we lost the Schiaparelli Mars Lander and Rover because of a software issue that caused the lander to believe it had already landed when it was actually nearly 4 kilometers above the surface.

In 1999, we lost the Mars Climate Orbiter because of a ground-based software error that transmitted data to the orbiter using English units (pound-seconds) rather than metric units (newton-seconds). This error caused the orbiter to descend too far into the Martian atmosphere, where it was torn to pieces.

In 1996, the Ariane 5 launch vehicle and payload was destroyed 37 seconds after launch because of an integer overflow exception when a 64-bit floating-point number underwent an unchecked conversion to a 16-bit integer. The exception crashed the onboard computers, and the vehicle self-destructed.

Should we talk about the Therac-25 radiation therapy machine that, because of a race condition, killed three people and injured three others by blasting them with a high-powered electron beam?

Or maybe we should talk about Knight Capital Group, which lost $460 million in 45 minutes because it reused a flag that activated dead code left in the system.

Or perhaps we should talk about the Toyota stack overflow bug that could cause cars to accelerate out of control—killing perhaps as many as 89 people.

Or maybe we should talk about HealthCare.gov, the software failure that nearly overturned a new and controversial American healthcare law.

These disasters have cost billions of dollars and many lives. And they were caused by programmers.

We, programmers, through the code that we write, are killing people.

Now, I know you didn't get into this business in order to kill people. Probably, you became a programmer because you wrote an infinite loop that printed your name once, and you experienced that joyous feeling of power.

But facts are facts. We are now in a position in our society where our actions can destroy fortunes, livelihoods, and lives.

One day, probably not too long from now, some poor programmer is going to do something just a little dumb, and tens of thousands of people will die.

This isn't wild speculation—it's just a matter of time.

And when this happens, the politicians of the world will demand an accounting. They will demand that we show them how we will prevent such errors in the future.

And if we show up without a statement of ethics, if we show up without any standards or defined disciplines, if we show up and whine about how our bosses set unreasonable schedules and deadlines—then we will be found GUILTY.

The Oath（誓言）

To begin the discussion of our ethics as software developers, I offer the following oath.

> *In order to defend and preserve the honor of the profession of computer programmers, I promise that, to the best of my ability and judgment:*
>
> 1. *I will not produce harmful code.*
> 2. *The code that I produce will always be my best work. I will not knowingly allow code that is defective either in behavior or structure to accumulate.*
> 3. *I will produce, with each release, a quick, sure, and repeatable proof that every element of the code works as it should.*
> 4. *I will make frequent, small releases so that I do not impede the progress of others.*
> 5. *I will fearlessly and relentlessly improve my creations at every opportunity. I will never degrade them.*
> 6. *I will do all that I can to keep the productivity of myself and others as high as possible. I will do nothing that decreases that productivity.*
> 7. *I will continuously ensure that others can cover for me and that I can cover for them.*
> 8. *I will produce estimates that are honest in both magnitude and precision. I will not make promises without reasonable certainty.*
> 9. *I will respect my fellow programmers for their ethics, standards, disciplines, and skill. No other attribute or characteristic will be a factor in my regard for my fellow programmers.*
> 10. *I will never stop learning and improving my craft.*

12 HARM

（伤害）

Several promises in the oath are related to harm.

First, Do No Harm（首先，不造成伤害）

Promise 1. I will not produce harmful code.

The first promise of the software professional is DO NO HARM! That means that your code must not harm your users, your employers, your managers, or your fellow programmers.

You must know what your code does. You must know that it works. And you must know that it is clean.

Some time back, it was discovered that some programmers at Volkswagen wrote some code that purposely thwarted Environmental Protection Agency (EPA) emissions tests. Those programmers wrote harmful code. It was harmful because it was deceitful. That code fooled the EPA into allowing cars to be sold that emitted twenty times the amount of harmful nitrous oxides that the EPA deemed safe. Therefore, that code potentially harmed the health of everyone living where those cars were driven.

What should happen to those programmers? Did they know the purpose of that code? Should they have known?

I'd fire them and prosecute them because, whether they knew or not, they *should* have known. Hiding behind requirements written by others is no excuse. It's your fingers on the keyboard, it's your code. You must know what it does!

That's a tough one, isn't it? We write the code that makes our machines work, and those machines are often in positions to do tremendous harm. Because we will be held responsible for any harm our code does, we must be responsible for knowing what our code will do.

Each programmer should be held accountable based on their level of experience and responsibility. As you advance in experience and position, your responsibility for your actions and the actions of those under your charge increases.

Clearly, we can't hold junior programmers as responsible as team leads. We can't hold team leads as responsible as senior developers. But those senior people ought to be held to a very high standard and be ultimately responsible for those whom they direct.

That doesn't mean that all the blame goes to the senior developers or the managers. Every programmer is responsible for knowing what the code does to the level of their maturity and understanding. Every programmer is responsible for the harm their code does.

No Harm to Society（对社会无害）

First, you will do no harm to the society in which you live.

This is the rule that the VW programmers broke. Their software might have benefited their employer—Volkswagen. However, it harmed society in general. And we, programmers, must never do that.

But how do you know whether or not you are harming society? For example, is building software that controls weapon systems harmful to society? What about gambling software? What about violent or sexist video games? What about pornography?

If your software is within the law, might it still be harmful to society?

Frankly, that's a matter for your own judgment. You'll just have to make the best call you can. Your conscience is going to have to be your guide.

Another example of harm to society was the failed launch of HealthCare.gov, although in this case the harm was unintentional. The Affordable Care Act was passed by the US Congress and signed into law by the president in 2010.

Among its many directives was the demand that a Web site be created and activated on October 1, 2013.

Never mind the insanity of specifying, by law, a date certain by which a whole new massive software system must be activated. The real problem was that on October 1, 2013, they actually turned it on.

Do you think, maybe, there were some programmers hiding under their desks that day?

Oh man, I think they turned it on.

Yeah, yeah, they really shouldn't be doing that.

Oh, my poor mother. Whatever will she do?

This is a case where a technical screwup put a huge new public policy at risk. The law was very nearly overturned because of these failures. And no matter what you think about the politics of the situation, that was harmful to society.

Who was responsible for that harm? Every programmer, team lead, manager, and director who knew that system wasn't ready and who nevertheless remained silent.

That harm to society was perpetrated by every software developer who maintained a passive-aggressive attitude toward their management, everyone who said, "I'm just doing my job—it's their problem." Every software developer who knew something was wrong and yet did nothing to stop the deployment of that system shares part of the blame.

Because here's the thing. One of the prime reasons you were hired as a programmer was because you know when things are going wrong. You have the knowledge to identify trouble before it happens. Therefore, you have the responsibility to speak up before something terrible happens.

Harm to Function（对功能的损害）

You must KNOW that your code works. You must KNOW that the functioning of your code will not do harm to your company, your users, or your fellow programmers.

On August 1, 2012, some technicians at Knight Capital Group loaded their servers with new software. Unfortunately, they loaded only seven of the eight servers, leaving the eighth running with the older version.

Why they made this mistake is anybody's guess. Somebody got sloppy.

Knight Capital ran a trading system. It traded stocks on the New York Stock Exchange. Part of its operation was to take large parent trades and break them up into many smaller child trades in order to prevent other traders from seeing the size of the initial parent trade and adjusting prices accordingly.

Eight years earlier, a simple version of this parent–child algorithm, named Power Peg, was disabled and replaced with something much better called Smart Market Access Routing System (SMARS). Oddly, however, the old Power Peg code was not removed from the system. It was simply disabled with a flag.

That flag had been used to regulate the parent–child process. When the flag was on, child trades were made. When enough child trades had been made to satisfy the parent trade, the flag was turned off.

They disabled the Power Peg code by simply leaving the flag off.

Unfortunately, the new software update that made it on to only seven of the eight servers repurposed that flag. The update turned the flag on, and that eighth server started making child trades in a high-speed infinite loop.

The programmers knew something had gone wrong, but they didn't know exactly what. It took them 45 minutes to shut down that errant server—45 minutes during which it was making bad trades in an infinite loop.

The bottom line is that, in that first 45 minutes of trading, Knight Capital had unintentionally purchased more than $7 billion worth of stock that it did not want and had to sell it off at a $460 million loss. Worse, the company only had $360 million in cash. Knight Capital was bankrupt.

Forty-five minutes. One dumb mistake. Four hundred sixty million dollars.

And what was that mistake? The mistake was that the programmers did not KNOW what their system was going to do.

At this point, you may be worried that I'm demanding that programmers have perfect knowledge about the behavior of their code. Of course, perfect knowledge is not achievable; there will always be a knowledge deficit of some kind.

The issue isn't the perfection of knowledge. Rather, the issue is to KNOW that there will be no harm.

Those poor guys at Knight Capital had a knowledge deficit that was horribly harmful—and given what was at stake, it was one they should not have allowed to exist.

Another example is the case of Toyota and the software system that made cars accelerate uncontrollably.

As many as 89 people were killed by that software, and more were injured.

Imagine that you are driving in a crowded downtown business district. Imagine that your car suddenly begins to accelerate, and your brakes stop working. Within seconds, you are rocketing through stoplights and crosswalks with no way to stop.

That's what investigators found that the Toyota software could do—and quite possibly had done.

The software was killing people.

The programmers who wrote that code did not KNOW that their code would not kill—notice the double negative. They did not KNOW that their code would NOT kill. And they should have known that. They should have known that their code would NOT kill.

Now again, this is all about risk. When the stakes are high, you want to drive your knowledge as near to perfection as you can get it. If lives are at stake, you have to KNOW that your code won't kill anyone. If fortunes are a stake, you have to KNOW that your code won't lose them.

On the other hand, if you are writing a chat application or a simple shopping cart Web site, neither fortunes nor lives are at stake . . .

. . . or are they?

What if someone using your chat application has a medical emergency and types "HELP ME. CALL 911" on your app? What if your app malfunctions and drops that message?

What if your Web site leaks personal information to hackers who use it to steal identities?

And what if the poor functioning of your code drives customers away from your employer and to a competitor?

The point is that it's easy to underestimate the harm that can be done by software. It's comforting to think that your software can't harm anyone because it's not important enough. But you forget that software is very expensive to write, and at very least, the money spent on its development is at stake—not to mention whatever the users put at stake by depending on it.

The bottom line is, there's almost always more at stake than you think.

No Harm to Structure（对结构无害）

You must not harm the structure of the code. You must keep the code clean and well organized.

Ask yourself why the programmers at Knight Capital did not KNOW that their code could be harmful.

I think the answer is pretty obvious. They had forgotten that the Power Peg software was still in the system. They had forgotten that the flag they repurposed would activate it. And they assumed that all servers would always have the same software running.

They did not know their system could exhibit harmful behavior because of the harm they had done to the structure of the system by leaving that dead code in place. And that's one of the big reasons why code structure and cleanliness are so important. The more tangled the structure, the more difficulty in knowing what the code will do. The more the mess, the more the uncertainty.

Take the Toyota case, for example. Why didn't the programmers know their software could kill people? Do you think the fact that they had more than ten thousand global variables might have been a factor?

Making a mess in the software undermines your ability to know what the software does and therefore your ability to prevent harm.

Messy software is harmful software.

Some of you might object by saying that sometimes a quick and dirty patch is necessary to fix a nasty production bug.

Sure. Of course. If you can fix a production crisis with a quick and dirty patch, then you should do it. No question.

A stupid idea that works is not a stupid idea.

However, you can't leave that quick and dirty patch in place without causing harm. The longer that patch remains in the code, the more harm it can do.

Remember that the Knight Capital debacle would not have happened if the old Power Peg code had been removed from the code base. It was that old defunct code that actually made the bad trades.

What do we mean by "harm to structure"? Clearly, having thousands of global variables is a structural flaw. So is dead code left in the code base.

Structural harm is harm to the organization and content of the source code. It is anything that makes that source code hard to read, hard to understand, hard to change, or hard to reuse.

It is the responsibility of every professional software developer to know the disciplines and standards of good software structure. They should know how to refactor, how to write tests, how to recognize bad code, how to decouple designs and create appropriate architectural boundaries. They should know and apply the principles of low- and high-level design. And it is the responsibility of every senior developer to make sure that younger developers learn these things and satisfy them in the code that they write.

SOFT（柔软）

The first word in software is *soft*. Software is supposed to be SOFT. It's supposed to be easy to change. If we didn't want it to be easy to change, we'd have called it *hard*ware.

It's important to remember why software exists at all. We invented it as a means to make the behavior of machines easy to change. To the extent that our software is hard to change, we have thwarted the very reason that software exists.

Remember that software has two values. There's the value of its behavior, and then there's the value of its "softness." Our customers and users expect us to be able to change that behavior easily and without high cost.

Which of these two values is the greater? Which value should we prioritize above the other? We can answer that question with a simple thought experiment.

Imagine two programs. One works perfectly but is impossible to change. The other does nothing correctly but is easy to change. Which is the more valuable?

I hate to be the one to tell you this but, in case you hadn't noticed, software requirements tend to change, and when they change, that first software will become useless—forever.

On the other hand, that second software can be made to work, because it's easy to change. It may take some time and money to get it to work initially, but after that, it'll continue to work forever with minimal effort.

Therefore, it is the second of the two values that should be given priority in all but the most urgent of situations.

What do I mean by *urgent?* I mean a production disaster that's losing the company $10 million per minute. That's urgent.

I do not mean a software startup. A startup is not an urgent situation requiring you to create inflexible software. Indeed, the opposite is true. The one thing that is absolutely certain in a startup is that you are creating the wrong product.

No product survives contact with the user. As soon as you start to put the product into users' hands, you will discover that the product you've built is wrong in a hundred different ways. And if you can't change it without making a mess, you're doomed.

This is one of the biggest problems with software startups. Startup entrepreneurs believe they are in an urgent situation requiring them to throw out all the rules and dash to the finish line, leaving a huge mess in their wake. Most of the time, that huge mess starts to slow them down long before they do their first deployment. They'll go faster and better and will survive a lot longer if they keep the structure of the software from harm.

> *When it comes to software, it never pays to rush.*
>
> —Brian Marick

TESTS（测试）

Tests come first. You write them first, and you clean them first. You will know that every line of code works because you will have written tests that prove they work.

How can you prevent harm to the behavior of your code if you don't have the tests that prove that it works?

How can you prevent harm to the structure of your code if you don't have the tests that allow you to clean it?

And how can you guarantee that your test suite is complete if you don't follow the three laws of test-driven development (TDD)?

Is TDD really a prerequisite to professionalism? Am I really suggesting that you can't be a professional software developer unless you practice TDD?

Yes, I think that's true. Or rather, it is becoming true. It is true for some of us, and with time, it is becoming true for more and more of us. I think the time will come, and relatively soon, when the majority of programmers will agree that practicing TDD is part of the minimum set of disciplines and behaviors that mark a professional developer.

Why do I believe that?

Because, as I said earlier, we rule the world! We write the rules that make the whole world work.

In our society, nothing gets bought or sold without software. Nearly all correspondence is through software. Nearly all documents are written with software. Laws don't get passed or enforced without it. There is virtually no activity of daily life that does not involve software.

Without software, our society does not function. Software has become the most important component in the infrastructure of our civilization.

Society does not understand this yet. We programmers don't really understand it yet either. But the realization is dawning that the software we write is critical. The realization is dawning that many lives and fortunes depend on our software. And the realization is dawning that software is being written by people who do not profess a minimum level of discipline.

So, yes, I think TDD, or some discipline very much like it, will eventually be considered a minimum standard behavior for professional software developers. I think our customers and our users will insist on it.

Best Work（最好的作品）

Promise 2. The code that I produce will always be my best work. I will not knowingly allow code that is defective in either behavior or structure to accumulate.

Kent Beck once said, "First make it work. Then make it right."

Getting the program to work is just the first—and easiest—step. The second—and harder—step is to clean the code.

Unfortunately, too many programmers think they are done once they get a program to work. Once it works, they move on to the next program, and then the next, and the next.

In their wake, they leave behind a history of tangled, unreadable, code that slows down the whole development team. They do this because they think their value is in speed. They know they are paid a lot, and so they feel that they must deliver a lot of functionality in a short amount of time.

But software is hard and takes a lot of time, and so they feel like they are going too slowly. They feel like they are failing, which creates a pressure that causes them to try to go faster. It causes them to rush. They rush to get the program working, and then they declare themselves to be done—because, in their minds, it's already taken them too long. The constant tapping of the project manager's foot doesn't help, but that's not the real driver.

I teach lots of classes. In many of them, I give the programmers small projects to code. My goal is to give them a coding experience in which to try out new techniques and new disciplines. I don't care if they actually finish the project. Indeed, all the code is going to be thrown away.

Yet still I see people rushing. Some stay past the end of the class just hammering away at getting something utterly meaningless to work.

So, although the boss's pressure doesn't help. The real pressure comes from inside us. We consider speed of development to be a matter of our own self-worth.

Making It Right（使其正确）

As we saw earlier in this chapter, there are two values to software. There is the value of its behavior and the value of its structure. I also made the point that the value of structure is more important than the value of behavior. This is because to have any long-term value, software systems must be able to react to changes in requirements.

Software that is hard to change also is hard to keep up to date with the requirements. Software with a bad structure can quickly get out of date.

In order for you to keep up with requirements, the structure of the software has to be clean enough to allow, and even encourage, change. Software that is easy to change can be kept up to date with changing requirements, allowing it to remain valuable with the least amount of effort. But if you've got a software system that's hard to change, then you're going to have a devil of a time keeping that system working when the requirements change.

When are requirements most likely to change? Requirements are most volatile at the start of a project, just after the users have seen the first few features work. That's because they're getting their first look at what the system actually does as opposed to what they thought it was going to do.

Consequently, the structure of the system needs to be clean at the very start if early development is to proceed quickly. If you make a mess at the start, then even the very first release will be slowed down by that mess.

Good structure enables good behavior and bad structure impedes good behavior. The better the structure, the better the behavior. The worse the structure, the worse the behavior. The value of the behavior depends critically on the structure. Therefore, the value of the structure is the most critical of the two values, which means that professional developers put a higher priority on the structure of the code than on the behavior.

Yes, first you make it work; but then you make very sure that you continue to make it right. You keep the system structure as clean as possible throughout the lifetime of the project. From its very beginning to its very end, it must be clean.

What Is Good Structure?（什么是好结构）

Good structure makes the system easy to test, easy to change, and easy to reuse. Changes to one part of the code do not break other parts of the code. Changes to one module do not force massive recompiles and redeployments. High-level policies are kept separate and independent from low-level details.

Poor structure makes a system rigid, fragile, and immobile. These are the traditional *design smells*.

Rigidity is when relatively minor changes to the system cause large portions of the system to be recompiled, rebuilt, and redeployed. A system is rigid when the effort to integrate a change is much greater than the change itself.

Fragility is when minor changes to the behavior of a system force many corresponding changes in a large number of modules. This creates a high risk that a small change in behavior will break some other behavior in the system. When this happens, your managers and customers will come to believe that you have lost control over the software and don't know what you are doing.

Immobility is when a module in an existing system contains a behavior you want to use in a new system but is so tangled up within the existing system that you can't extract it for use in the new system.

These problems are all problems of structure, not behavior. The system may pass all its tests and meet all its functional requirements, yet such a system may be close to worthless because it is too difficult to manipulate.

There's a certain irony in the fact that so many systems that correctly implement valuable behaviors end up with a structure that is so poor that it negates that value and makes the system worthless.

And *worthless* is not too strong a word. Have you ever participated in the *Grand Redesign in the Sky?* This is when developers tell management that the only way to make progress is to redesign the whole system from scratch; those developers have assessed the current system to be worthless.

When managers agree to let the developers redesign the system, it simply means that they have agreed with the developers' assessment that the current system is worthless.

What is it that causes these design smells that lead to worthless systems? Source code dependencies! How do we fix those dependencies? Dependency management!

How do we manage dependencies? We use the SOLID[5] principles of object-oriented design to keep the structure of our systems free of the design smells that would make it worthless.

Because the value of structure is greater than the value of behavior and depends on good dependency management, and because good dependency

5. These principles are described in Robert C. Martin, *Clean Code* (Addison-Wesley, 2009), and *Agile Software Development: Principles, Patterns, and Practices* (Pearson, 2003).

management derives from the SOLID principles, it follows that overall value of the system depends on proper application of the SOLID principles.

That's quite a claim, isn't it? Perhaps it's a bit hard to believe. The value of the system depends on design principles. But we've gone through the logic, and many of you have had the experiences to back it up. So, that conclusion bears some serious consideration.

Eisenhower's Matrix（艾森豪威尔矩阵）

General Dwight D. Eisenhower once said, "I have two kinds of problems, the urgent and the important. The urgent are not important, and the important are never urgent."

There is a deep truth to this statement—a deep truth about engineering. We might even call this the engineer's motto:

> *The greater the urgency, the less the relevance.*

Figure 12.1 presents Eisenhower's decision matrix: urgency on the vertical axis, importance on the horizontal. The four possibilities are urgent and important, urgent and unimportant, important but not urgent, and neither important nor urgent.

IMPORTANT URGENT	IMPORTANT NOT URGENT
UNIMPORTANT URGENT	UNIMPORTANT NOT URGENT

Figure 12.1 Eisenhower's decision matrix

Now let's arrange these in order of priority. The two obvious cases are important and urgent at the top and neither important nor urgent at the bottom.

The question is, how do you sort the two in the middle: urgent but unimportant and important but not urgent? Which should you address first?

Clearly, things that are important should be prioritized over things that are unimportant. I would argue, furthermore, that if something is unimportant, it shouldn't be done at all. Doing unimportant things is a waste.

If we eliminate all the unimportant things, we have two left. We do the important and urgent things first. Then we do the important but not urgent things second.

My point is that urgency is about time. Importance is not. Things that are important are long-term. Things that are urgent are short-term. Structure is long-term. Therefore, it is important. Behavior is short-term. Therefore, it is merely urgent.

So, structure, the important stuff, comes first. Behavior is secondary.

Your boss might not agree with that priority, but that's because it's not your boss's job to worry about structure. It's yours. Your boss simply expects that you will keep the structure clean while you implement the urgent behaviors.

Earlier in this chapter, I quoted Kent Beck: "First make it work, then make it right." Now I'm saying that structure is a higher priority than behavior. It's a chicken-and-egg dilemma, isn't it?

The reason we make it work first is that structure has to support behavior; consequently, we implement behavior first, then we give it the right structure. But structure is more important than behavior. We give it a higher priority. We deal with structural issues before we deal with behavioral issues.

We can square this circle by breaking the problem down into tiny units. Let's start with user stories. You get a story to work, and then you get its structure right. You don't work on the next story *until* that structure is right. The structure of the current story is a higher priority than the behavior of the next story.

Except, stories are too big. We need to go smaller. Not stories, then—tests. Tests are the perfect size.

First you get a test to pass, then you fix the structure of the code that passes that test before you get the next test to pass.

This entire discussion has been the *moral* foundation for the red → green → refactor cycle of TDD.

It is that cycle that helps us to prevent harm to behavior and harm to structure. It is that cycle that allows us to prioritize structure over behavior. And that's why we consider TDD to be a *design* technique as opposed to a testing technique.

PROGRAMMERS ARE STAKEHOLDERS（程序员是利益相关者）

Remember this. We have a stake in the success of the software. We, programmers, are stakeholders too.

Did you ever think about it that way? Did you ever view yourself as one of the stakeholders of the project?

But, of course, you are. The success of the project has a direct impact on your career and reputation. So, yes, you are a stakeholder.

And as a stakeholder, you have a say in the way the system is developed and structured. I mean, it's your butt on the line too.

But you are more than just a stakeholder. You are an engineer. You were hired because *you* know how to build software systems and how to structure those systems so that they last. Along with that knowledge comes the responsibility to produce the best product you can.

Not only do you have the right as a stakeholder, but you have the duty as an engineer, to make sure that the systems you produce do no harm either through bad behavior or bad structure.

A lot of programmers don't want that kind of responsibility. They'd rather just be told what to do. And that's a travesty and a shame. It's wholly unprofessional. Programmers who feel that way should be paid minimum wage because that's what their work output is worth.

If you don't take responsibility for the structure of the system, who else will? Your boss?

Does your boss know the SOLID principles? How about design patterns? How about object-oriented design and the practice of dependency inversion? Does your boss know the discipline of TDD? What a self-shunt is, or a test-specific subclass, or a Humble Object? Does your boss understand that things that change together should be grouped together and that things that change for different reasons should be separated?

Does your boss understand structure? Or is your boss's understanding limited to behavior?

Structure matters. If you aren't going to care for it, who will?

What if your boss specifically tells you to ignore structure and focus entirely on behavior? You refuse. You are a stakeholder. You have rights too. You are also an engineer with responsibilities that your boss cannot override.

Perhaps you think that refusing will get you fired. It probably will not. Most managers expect to have to fight for things they need and believe in, and they respect those who are willing to do the same.

Oh, there will be a struggle, even a confrontation, and it won't be comfortable. But you are a stakeholder and an engineer. You can't just back down and acquiesce. That's not professional.

Most programmers do not enjoy confrontation. But dealing with confrontational managers is a skill we have to learn. We have to learn how to fight for what we know is right because taking responsibility for the things that matter, and fighting for those things, is how a professional behaves.

YOUR BEST（尽力而为）

This promise of the Programmer's Oath is about doing your best.

Clearly, this is a perfectly reasonable promise for a programmer to make. Of course you are going to do your best, and of course you will not knowingly release code that is harmful.

And, of course, this promise is not always black and white. There are times when structure must bend to schedule. For example, if in order to make a trade show, you have to put in some quick and dirty fix, then so be it.

The promise doesn't even prevent you from shipping to customers code with less-than-perfect structure. If the structure is close but not quite right, and customers are expecting the release tomorrow, then so be it.

On the other hand, the promise *does* mean that you will address those issues of behavior and structure before you add more behavior. You will not pile more and more behavior on top of known bad structure. You will not allow those defects to *accumulate*.

What if your boss tells you to do it anyway? Here's how that conversation should go.

Boss: I want this new feature added by tonight.

Programmer: *I'm sorry, but I can't do that. I've got some structural cleanup to do before I can add that new feature.*

Boss: *Do the cleanup tomorrow. Get the feature done by tonight.*

Programmer: *That's what I did for the last feature, and now I have an even bigger mess to clean up. I really have to finish that cleanup before I start on anything new.*

Boss: *I don't think you understand. This is business. Either we have a business or we don't have a business. And if we can't get features done, we don't have a business. Now get the feature done.*

Programmer: *I understand. Really, I do. And I agree. We have to be able to get features done. But if I don't clean up the structural problems that have accumulated over the last few days, we're going to slow down and get even fewer features done.*

Boss: *Ya know, I used to like you. I used to say, that Danny, he's pretty nice. But now I don't think so. You're not being nice at all. Maybe you shouldn't be working with me. Maybe I should fire you.*

Programmer: *Well, that's your right. But I'm pretty sure you want features done quickly and done right. And I'm telling you that if I don't do this cleanup tonight, then we're going to start slowing down. And we'll deliver fewer and fewer features.*

Look, I want to go fast, just like you do. You hired me because I know how to do that. You have to let me do my job. You have to let me do what I know is best.

Boss: *You really think everything will slow down if you don't do this cleanup tonight?*

Programmer: *I know it will. I've seen it before. And so have you.*

Boss: *And it has to be tonight?*

Programmer: *I don't feel safe letting the mess get any worse.*

Boss: *You can give me the feature tomorrow?*

Programmer: *Yes, and it'll be a lot easier to do once the structure is cleaned up.*

Boss: *Okay. Tomorrow. No later. Now get to it.*

Programmer: *Okay. I'll get right on it.*

Boss: [aside] I like that kid. He's got guts. He's got gumption. He didn't back down even when I threatened to fire him. He's gonna go far, trust me—but don't tell him I said so.

Repeatable Proof（可重复证据）

Promise 3. I will produce, with each release, a quick, sure, and repeatable proof that every element of the code works as it should.

Does that sound unreasonable to you? Does it sound unreasonable to be expected to prove that the code you've written actually works?

Allow me to introduce you to Edsger Wybe Dijkstra.

Dijkstra（狄克斯特拉）

Edsger Wybe Dijkstra was born in Rotterdam in 1930. He survived the bombing of Rotterdam and the German occupation of the Netherlands, and in 1948, he graduated high school with the highest possible marks in math, physics, chemistry, and biology.

In March 1952, at the age of 21, and just 9 months before I would be born, he took a job with the Mathematical Center in Amsterdam as the Netherlands' very first programmer.

In 1957, he married Maria Debets. In the Netherlands, at the time, you had to state your profession as part of the marriage rites. The authorities were unwilling to accept "programmer" as his profession. They'd never heard of such a profession. Dijkstra settled for "theoretical physicist."

In 1955, having been a programmer for 3 years, and while still a student, he concluded that the intellectual challenge of programming was greater than the intellectual challenge of theoretical physics, and as a result, he chose programming as his long-term career.

In making this decision, he conferred with his boss, Adriaan van Wijngaarden. Dijkstra was concerned that no one had identified a discipline or science of programming and that he would therefore not be taken seriously. His boss replied that Dijkstra might very well be one of the people who would make it a science.

In pursuit of that goal, Dijkstra was compelled by the idea that software was a formal system, a kind of mathematics. He reasoned that software could become a mathematical structure rather like *Euclid's Elements*—a system of postulates, proofs, theorems, and lemmas. He therefore set about creating the language and discipline of software proofs.

Proving Correctness（正确性证明）

Dijkstra realized that there were only three techniques we could use to prove the correctness of an algorithm: enumeration, induction, and abstraction. Enumeration is used to prove that two statements in sequence or two statements selected by a Boolean expression are correct. Induction is used to prove that a loop is correct. Abstraction is used to break groups of statements into smaller provable chunks.

If this sounds hard, it is.

As an example of just how hard this is, I have included a simple Java program for calculating the remainder of an integer (Figure 12.2), and the handwritten proof of that algorithm (Figure 12.3).[6]

```
public static int remainder(int numerator, int denominator) (
  assert(numerator > 0 && denominator > 0);
  int r = numerator;
  int dd = denominator;
  while(dd<=r)
    dd *= 2;
  while(dd != denominator) {
    dd /= 2;
    if(dd <= r)
      r -= dd;
  }
  return r;
}
```

Figure 12.2 A simple Java program

6. This is a translation into Java of a demonstration from Dijkstra's work.

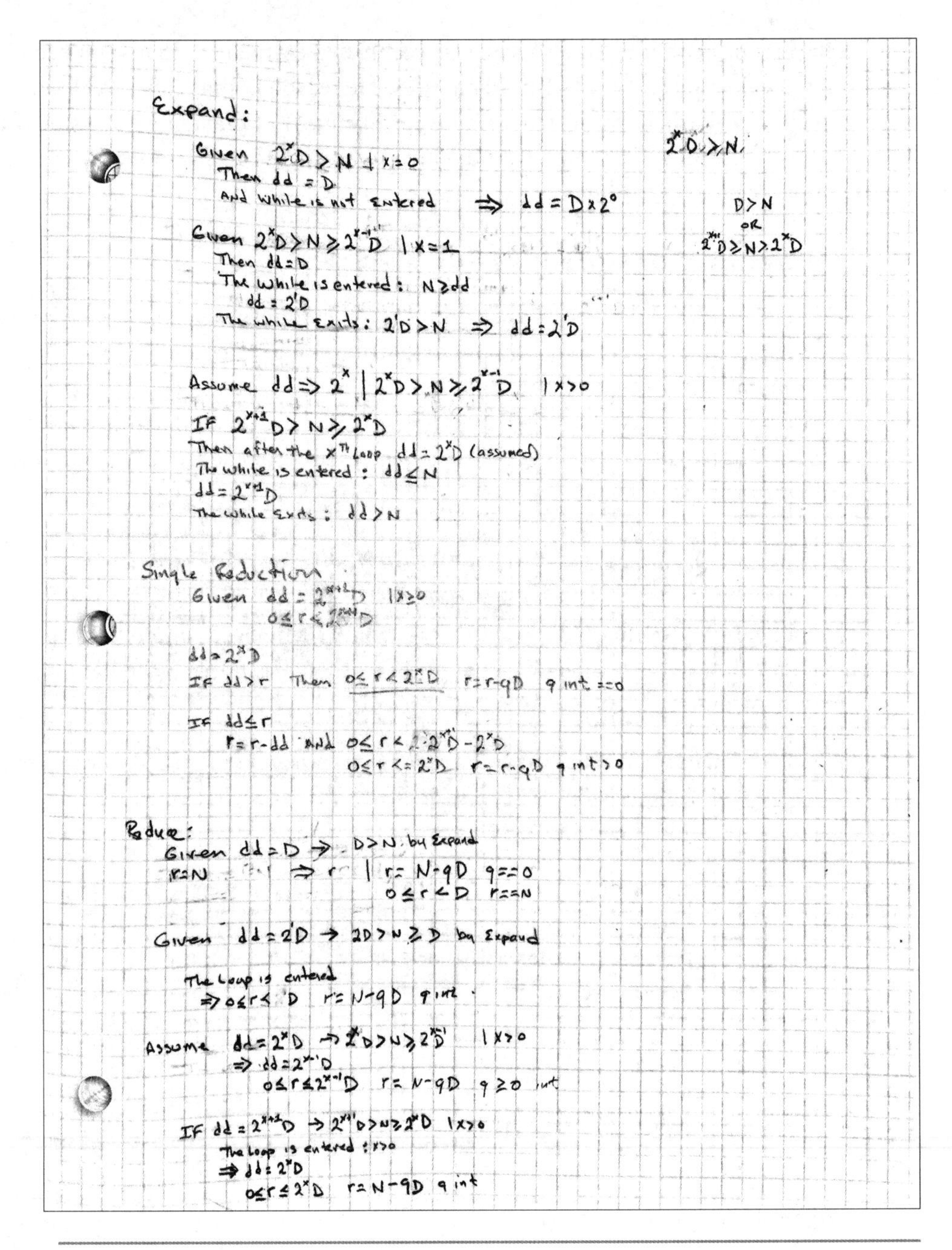

Figure 12.3 Handwritten proof of the algorithm

I think you can see the problem with this approach. Indeed, this is something that Dijkstra complained bitterly about:

> *Of course I would not dare to suggest (at least at present!) that it is the programmer's duty to supply such a proof whenever he writes a simple loop in his program. If so, he could never write a program of any size at all.*

Dijkstra's hope was that such proofs would become more practical through the creation of a library of theorems, again similar to *Euclid's Elements*.

But Dijkstra did not understand just how prevalent and pervasive software would become. He did not foresee, in those early days, that computers would outnumber people and that vast quantities of software would be running in the walls of our homes, in our pockets, and on our wrists. Had he known, he would have realized that the library of theorems he envisioned would be far too vast for any mere human to grasp.

So, Dijkstra's dream of explicit mathematical proofs for programs has faded into near oblivion. Oh, there are some holdouts who hope against hope for a resurgence of formal proofs, but their vision has not penetrated very far into the software industry at large.

Although the dream may have passed, it drew something deeply profound in its wake. Something that we use today, almost without thinking about it.

Structured Programming（结构化编程）

In the early days of programming, the 1950s and 1960s, we used languages such as Fortran. Have you ever seen Fortran? Here, let me show you what it was like.

```
        WRITE(4,99)
99      FORMAT(" NUMERATOR:")
        READ(4,100)NN
        WRITE(4,98)
98      FORMAT(" DENOMINATOR:")
        READ(4,100)ND
```

```
100     FORMAT(I6)
        NR=NN
        NDD=ND
1       IF(NDD-NR)2,2,3
2       NDD=NDD*2
        GOTO 1

3       IF(NDD-ND)4,10,4
4       NDD=NDD/2
        IF(NDD-NR)5,5,6
5       NR=NR-NDD
6       GOTO 3

10      WRITE(4,20)NR
20      FORMAT(" REMAINDER:",I6)
        END
```

This little Fortran program implements the same remainder algorithm as the earlier Java program.

Now I'd like to draw your attention to those `GOTO` statements. You probably haven't seen statements like that very often. The reason you haven't seen statements like that very often is that nowadays we look on them with disfavor. In fact, most modern languages don't even have `GOTO` statements like that anymore.

Why don't we favor `GOTO` statements? Why don't our languages support them anymore? Because in 1968, Dijkstra wrote a letter to the editor of *Communications of the ACM*, titled "Go To Statement Considered Harmful."[7]

Why did Dijkstra consider the `GOTO` statement to be harmful? It all comes back to the three strategies for proving a function correct: enumeration, induction, and abstraction.

7. Edsger W. Dijkstra, "Go To Statement Considered Harmful," *Communications of the ACM* 11, no. 3 (1968), 147–148.

Enumeration depends on the fact that each statement in sequence can be analyzed independently and that the result of one statement feeds into the next. It should be clear to you that in order for enumeration to be an effective technique for proving the correctness of a function, every statement that is enumerated must have a single entry point and a single exit point. Otherwise, we could not be sure of either the inputs or the outputs of a statement.

What's more, induction is simply a special form of enumeration, where we assume the enumerated statement is true for some x and then prove by enumeration that it is true for $x + 1$.

Thus, the body of a loop must be enumerable. It must have a single entry and a single exit.

`GOTO` is considered harmful because a `GOTO` statement can jump into or out of the middle of an enumerated sequence. `GOTO`s make enumeration intractable, making it impossible to prove an algorithm correct by enumeration or induction.

Dijkstra recommended that, in order to keep code provable, it be constructed of three standard building blocks.

- **Sequence,** which we depict as two or more statements ordered in time. This simply represents nonbranching lines of code.
- **Selection,** depicted as two or more statements selected by a predicate. This simply represents `if/else` and `switch/case` statements.
- **Iteration,** depicted as a statement repeated under the control of a predicate. This represents a `while` or `for` loop.

Dijkstra showed that any program, no matter how complicated, can be composed of nothing more than these three structures and that programs structured in that manner are provable.

He called the technique *structured programming*.

Why is this important if we aren't going to write those proofs? If something is provable, it means you can reason about it. If something is unprovable, it means you cannot reason about it. And if you can't reason about it, you can't properly test it.

Functional Decomposition（功能分解）

In 1968, Dijkstra's ideas were not immediately popular. Most of us were using languages that depended on `GOTO`, so the idea of abandoning `GOTO` or imposing discipline on `GOTO` was abhorrent.

The debate over Dijkstra's ideas raged for several years. We didn't have an Internet in those days, so we didn't use Facebook memes or flame wars. But we did write letters to the editors of the major software journals of the day. And those letters raged. Some claimed Dijkstra to be a god. Others claimed him to be a fool. Just like social media today, except slower.

But in time, the debate slowed, and Dijkstra's position gained increasing support until, nowadays, most of the languages we use simply don't have a `GOTO`.

Nowadays, we are all structured programmers because our languages don't give us a choice. We all build our programs out of sequence, selection, and iteration. And very few of us make regular use of unconstrained `GOTO` statements.

An unintended side effect of composing programs from those three structures was a technique called *functional decomposition*. Functional decomposition is the process whereby you start at the top level of your program and recursively break it down into smaller and smaller provable units. It is the reasoning process behind structured programming. Structured programmers reason from the top down through this recursive decomposition into smaller and smaller provable functions.

This connection between structured programming and functional decomposition was the basis for the structured revolution that took place in the 1970s and 1980s. People like Ed Yourdon, Larry Constantine, Tom DeMarco, and Meilir Page-Jones popularized the techniques of structured analysis and structured design during those decades.

Test-Driven Development（TDD）

TDD, the red → green → refactor cycle, is functional decomposition. After all, you have to write tests against small bits of the problem. That means that you must functionally decompose the problem into testable elements.

The result is that every system built with TDD is built from functionally decomposed elements that conform to structured programming. And that means the system they compose is provable.

And the tests are the proof.

Or, rather, the tests are the *theory*.

The tests created by TDD are not a formal, mathematical proof, as Dijkstra wanted. In fact, Dijkstra is famous for saying that tests can only prove a program wrong; they can never prove a program right.

This is where Dijkstra missed it, in my opinion. Dijkstra thought of software as a kind of mathematics. He wanted us to build a superstructure of postulates, theorems, corollaries, and lemmas.

Instead, what we have realized is that software is a kind of *science*. We validate that science with experiments. We build a superstructure of theories based on passing tests, just as all other sciences do.

Have we proven the theory of evolution, or the theory of relativity, or the Big Bang theory, or any of the major theories of science? No. We can't prove them in any mathematical sense.

But we believe them, within limits, nonetheless. Indeed, every time you get into a car or an airplane, you are betting your life that Newton's laws of motion are correct. Every time you use a GPS system, you are betting that Einstein's theory of relativity is correct.

The fact that we have not mathematically proven these theories correct does not mean that we don't have sufficient proof to depend on them, even with our lives.

That's the kind of proof that TDD gives us. Not formal mathematical proof but experimental empirical proof. The kind of proof we depend on every day.

And that brings us back to the third promise in the Programmer's Oath:

> *I will produce, with each release, a quick, sure, and repeatable proof that every element of the code works as it should.*

Quick, sure, and repeatable. *Quick* means that the test suite should run in a very short amount of time. Minutes instead of hours.

Sure means that when the test suite passes, you know you can ship.

Repeatable means that those tests can be run by anybody at any time to ensure that the system is working properly. Indeed, we want the tests run many times per day.

Some may think that it is too much to ask that programmers supply this level of proof. Some may think that programmers should not be held to this high a standard. I, on the other hand, can imagine no other standard that makes any sense.

When a customer pays us to develop software for them, aren't we honor bound to prove, to the best of our ability, that the software we've created does what that customer has paid us for?

Of course we are. We owe this promise to our customers, and our employers, and our teammates. We owe it to our business analysts, our testers, and our project managers. But mostly we owe this promise to ourselves. For how can we consider ourselves professionals if we cannot prove that the work we have done is the work we have been paid to do?

What you owe, when you make that promise, is not the formal mathematical proof that Dijkstra dreamed of; rather, it is the scientific suite of tests that covers all the required behavior, runs in seconds or minutes, and produces the same clear pass/fail result every time it is run.

13 INTEGRITY

（集成）

Several promises in the oath involve integrity.

SMALL CYCLES（小周期）

> *Promise 4. I will make frequent, small releases so that I do not impede the progress of others.*

Making small releases just means changing a small amount of code for each release. The system may be large, but the incremental changes to that system are small.

THE HISTORY OF SOURCE CODE CONTROL（源代码控制的历史）

Let's go back to the 1960s for a moment. What is your source code control system when your source code is punched on a deck of cards (Figure 13.1)?

Figure 13.1 Punch card

The source code is not stored on disk. It's not "in the computer." Your source code is, literally, in your hand.

What is the source code control system? It is your desk drawer.

When you literally *possess* the source code, there is no need to "control" it. Nobody else can touch it.

And this was the situation throughout much of the 1950s and 1960s. Nobody even dreamed of something like a source code control system. You simply kept the source code under control by putting it in a drawer or a cabinet.

If anyone wanted to "check out" the source code, they simply went to the cabinet and took it. When they were done, they put it back.

We certainly didn't have merge problems. It was physically impossible for two programmers to be making changes to the same modules at the same time.

But things started to change in the 1970s. The idea of storing your source code on magnetic tape or even on disk was becoming attractive. We wrote line editing programs that allowed us to add, replace, and delete lines in source files on tape. These programs weren't screen editors. We *punched* our add, change, and delete directives on cards. The editor would read the source tape, apply the changes from the edit deck, and write the new source tape.

You may think this sounds awful. Looking back on it—it was. But it was better than trying to manage programs on cards! I mean 6,000 lines of code on cards weighs 30 pounds. What would you do if you accidentally dropped those cards and watched as they spread all over the floor and under furniture and down into heating vents?

If you drop a tape, you can just pick it up again.

Anyway, notice what happened. We started with one source tape, and in the editing process, we wound up with a second, new source tape. But the old tape still existed. If we put that old tape back on the rack, someone else might inadvertently apply their own changes to it, leaving us with a merge problem.

To prevent that, we simply kept the master source tape in our possession until we were done with our edits and our tests. Then we put a new master source tape back on the rack. We controlled the source code by keeping possession of the tape.

Protecting our source code required a process and convention. A true source code control process had to be used. No software yet, just human rules. But still, the concept of source code control had become separate from the source code itself.

As systems became increasingly larger, they needed greater numbers of programmers working on the same code at the same time. Grabbing the master tape and holding it became a real nuisance for everyone else. I mean, you could keep the master tape out of circulation for a couple of days or more.

So, we decided to extract modules from the master tape. The whole idea of modular programming was pretty new at the time. The notion that a program could be made up of many different source files was revolutionary.

We therefore started using bulletin boards like the one in Figure 13.2.

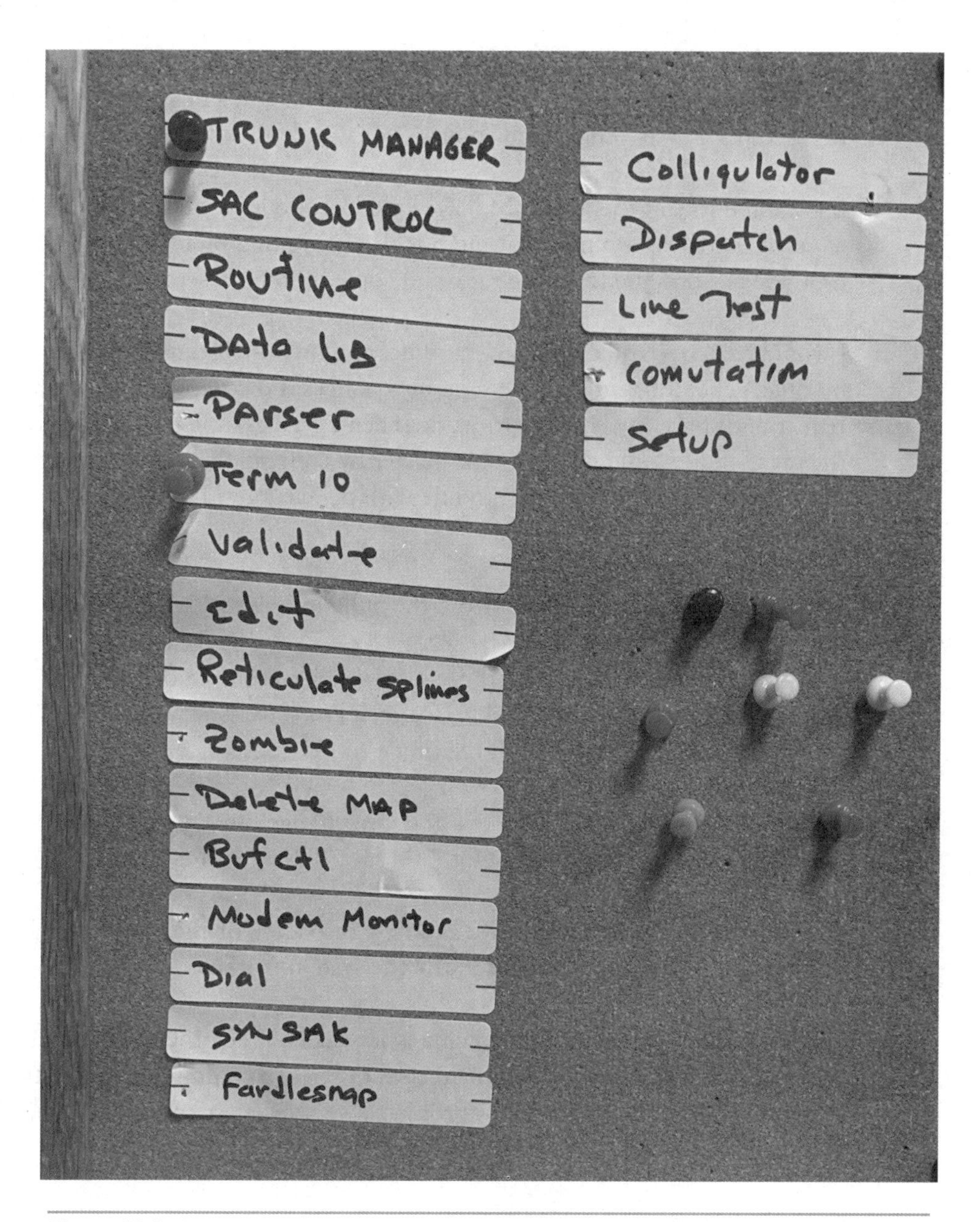

Figure 13.2 A bulletin board

The bulletin board had labels on it for each of the modules in the system. We programmers each had our own color thumbtack. I was blue. Ken was red. My buddy CK was yellow, and so on.

If I wanted to edit the Trunk Manager module, I'd look on the bulletin board to see if there was a pin in that module. If not, I put a blue pin in it. Then I took the master tape from the rack and copied it onto a separate tape.

I would edit the Trunk Manager module, and only the Trunk Manager module, generating a new tape with my changes. I'd compile, test, wash and repeat until I got my changes to work. Then I'd go get the master tape from the rack, and I would create a new master by copying the current master but replacing the Trunk Manager module with my changes. I'd put the new master back in the rack.

Finally, I'd remove my blue pin from the bulletin board.

This worked, but it only worked because we all knew each other, we all worked in the same office together, we each knew what the others were doing. And we *talked* to each other all the time.

I'd holler across the lab: "Ken, I'm going to change the Trunk Manager module." He'd say: "Put a pin in it." I'd say: "I already did."

The pins were just reminders. We all knew the status of the code and who was working on what. And that's why the system worked.

Indeed, it worked very well. Knowing what each of the other programmers was doing meant that we could help each other. We could make suggestions. We could warn each other about problems we'd recently encountered. And we could avoid merges.

Back then, merges were NOT fun.

Then, in the 1980s, came the disks. Disks got big, and they got permanent. By *big,* I mean hundreds of megabytes. By *permanent,* I mean that they were permanently mounted—always on line.

The other thing that happened is that we got machines like PDP11s and VAXes. We got screen editors, real operating systems, and multiple terminals. More than one person could be editing at exactly the same time.

The era of thumbtacks and bulletin boards had to come to an end.

First of all, by then, we had twenty or thirty programmers. There weren't enough thumbtack colors. Second, we had hundreds and hundreds of modules. We didn't have enough space on the bulletin board.

Fortunately, there was a solution.

In 1972, Marc Rochkind wrote the first source code control program. It was called the Source Code Control System (SCCS), and he wrote it in SNOBOL.[1]

Later, he rewrote it in C, and it became part of the UNIX distribution on PDP11s. SCCS worked on only one file at a time but allowed you to lock that file so that no one else could edit it until you were done. It was a life saver.

In 1982, RCS, the Revision Control System, was created by Walter Tichy. It too was file based and knew nothing of projects, but it was considered an improvement over SCCS and rapidly became the standard source code control system of the day.

Then, in 1986, CVS, the Concurrent Versions System, came along. It extended RCS to deal with whole projects, not just individual files. It also introduced the concept of *optimistic locking.*

1. A lovely little string processing language from the 1960s that had many of the pattern-matching facilities of our modern languages.

Up to this time, most source code control systems worked like my thumbtacks. If you had checked out the module, nobody else could edit it. This is called *pessimistic locking*.

CVS used optimistic locking. Two programmers could check out and change the same file at the same time. CVS would try to merge any nonconflicting changes and would alert you if it couldn't figure out how to do the merge.

After that, source code control systems exploded and even became commercial products. Literally hundreds of them flooded the market. Some used optimistic locking, others used pessimistic locking. The locking strategy became a kind of religious divide in the industry.

Then, in 2000, Subversion was created. It vastly improved upon CVS, and was instrumental in driving the industry away from pessimistic locking once and for all. Subversion was also the first source code control system to be used in the cloud. Does anybody remember Source Forge?

Up to this point, all source code control systems were based on the master tape idea that I used back in my bulletin board days. The source code was maintained in a single central master repository. Source code was checked out from that master repository, and commits were made back into that master repository.

But all that was about to change.

Git

The year is 2005. Multi-gigabyte disks are in our laptops. Network speeds are fast and getting faster. Processor clock rates have plateaued at 2.6GHz.

We are very, very far beyond my old bulletin board control system for source code. But we are still using the master tape concept. We still have a central repository that everybody has to check in and out of. Every commit, every reversion, every merge requires network connectivity to the master.

And then came git.

Well, actually, git was presaged by BitKeeper and monotone; but it was git that caught the attention of the programming world and changed everything.

Because, you see, git eliminates the master tape.

Oh, you still need a final authoritative version of the source code. But git does not automatically provide this location for you. Git simply doesn't care about that. Where you decide to put your authoritative version is entirely up to you. Git has nothing whatever to do with it.

Git keeps the entire history of the source code on your local machine. Your laptop, if that's what you use. On your machine, you can commit changes, create branches, check out old versions, and generally do anything you could do with a centralized system like Subversion—except you don't need to be connected to some central server.

Any time you like, you can connect to another user and push any of the changes you've made to that user. Or you can pull any changes they have made into your local repository. Neither is master. Both are equal. That's why they call it peer to peer.

And the final authoritative location that you use for making production releases is just another peer that people can push to or pull from any time they like.

The end result is that you are free to make as many small commits as you like before you push your changes somewhere else. You can commit every 30 seconds if you like. You can make a commit every time you get a unit test to pass.

And that brings us to the point of this whole historical discussion.

If you stand back and look at the trajectory of the evolution of source code control systems, you can see that they have been driven, perhaps unconsciously, by a single underlying imperative.

SHORT CYCLES（短周期）

Consider again how we began. How long was the cycle when source code was controlled by physically possessing a deck of cards?

You checked the source code out by taking those cards out of the cabinet. You held on to those cards until you were done with your project. Then you committed your changes by putting the changed deck of cards back in the cabinet. The cycle time was the whole project.

Then, when we used thumbtacks on the bulletin board, the same rule applied. You kept your thumbtacks in the modules you were changing until you were done with the project you were working on.

Even in the late 1970s and into the 1980s when we were using SCCS and RCS, we continued to use this pessimistic locking strategy, keeping others from touching the modules we were changing until we were done.

But CVS changed things—at least for some of us. Optimistic locking meant that one programmer could not lock others out of a module. We still committed only when we were done with a project, but others could be working concurrently on the same modules. Consequently, the average time between commits on a project shrank drastically. The cost of concurrency, of course, was merges.

And how we hated doing merges. Merges are awful. Especially without unit tests! They are tedious, time-consuming, and dangerous.

Our distaste for merges drove us to a new strategy.

Continuous Integration（持续集成）

By 2000, even while we were using tools such as Subversion, we had begun teaching the discipline of committing every few minutes.

The rationale was simple. The more frequently you commit, the less likely you are to face a merge. And if you do have to merge, the merge will be trivial.

We called this discipline *continuous integration*.

Of course, continuous integration depends critically on having a very reliable suite of unit tests. I mean, without good unit tests, it would be easy to make a merge error and break someone else's code. So, continuous integration goes hand in hand with test-driven development (TDD).

With tools like git, there is almost no limit to how small we can shrink the cycle. And that begs the question: Why are we so concerned about shrinking the cycle?

Because long cycles impede the progress of the team.

The longer the time between commits, the greater the chance that someone else on the team—perhaps the whole team—will have to wait for you. And that violates the promise.

Perhaps you think this is only about production releases. No, it's actually about every other cycle. It's about iterations and sprints. It's about the edit/compile/test cycle. It's about the time between commits. It's about *everything*.

And remember the rationale: so that you do not impede the progress of others.

Branches versus Toggles（分支与切换）

I used to be a branch nazi. Back when I was using CVS and Subversion, I refused to allow members on my teams to branch the code. I wanted all changes returned to the main line as frequently as possible.

My rationale was simple. A branch is simply a long-term checkout. And, as we've seen, long-term checkouts impede the progress of others by prolonging the time between integrations.

But then I switched to git—and everything changed overnight.

At the time, I was managing the open source FitNesse project, which had a dozen or so people working on it. I had just moved the FitNesse repository from Subversion (Source Forge) to git (GitHub). Suddenly, branches started appearing all over the place.

For the first few days, these crazy branches in git had me confused. Should I abandon my branch-nazi stance? Should I abandon continuous integration and simply allow everyone to make branches willy-nilly, forgetting about the cycle time issue?

But then it occurred to me that these branches I was seeing were not true named branches. Instead, they were just the stream of commits made by a developer between pushes. In fact, all git had really done was record the actions of the developer between continuous integration cycles.

So, I resolved to continue my rule to restrict branches. It's just that now, it's not commits that return immediately to the mainline. It's pushes. Continuous integration was preserved.

If we follow continuous integration and push to the main line every hour or so, then we'll certainly have a bunch of half-written features on the main line. There are typically two strategies for dealing with that: branches and toggles.

The branching strategy is simple. You simply create a new branch of the source code in which to develop the feature. You merge it back when the feature is done. This is most often accomplished by delaying the push until the feature is complete.

If you keep the branch out of the main line for days or weeks, then you'll likely face a big merge, and you'll certainly be impeding the team.

However, there are cases in which the new feature is so isolated from the rest of the code that branching it is not likely to cause a big merge. In these circumstances, it might be better to let the developers work in peace on the new feature without continuously integrating.

In fact, we had a case like this with FitNesse a few years ago. We completely rewrote the parser. This was a big project. It took a few man weeks. And there was no way to do it incrementally. I mean, the parser is the parser.

We therefore created a branch and kept that branch isolated from the rest of the system until the parser was ready.

There was a merge to do at the end, but it wasn't too bad. The parser was isolated well enough from the rest of the system. And, fortunately, we had a very comprehensive set of unit and acceptance tests.

Despite the success of the parser branch, I think it's usually better to keep new feature development on the main line and use toggles to turn those features off until ready.

Sometimes we use flags for those toggles, but more often we use the *Command* pattern, the *Decorator* pattern, and special versions of the *Factory* pattern to make sure that the partially written features cannot be executed in a production environment.

And most of the time we simply don't give the user the option to use the new feature. I mean, if the button isn't on the Web page, you can't execute that feature.

In many cases, of course, new features will be completed as part of the current iteration—or at least before the next production release—so there's no real need for any kind of toggle.

You only need a toggle if you are going to be releasing to production while some of the features are unfinished. How often should that be?

Continuous Deployment（持续部署）

What if we could eliminate the delays between production releases? What if we could release to production several times per day? After all, delaying production releases impedes others.

I want you to be able to release your code to production several times per day. I want you to feel comfortable enough with your work that you could release your code to production on every push.

This, of course, depends on automated testing: automated tests written by programmers to cover every line of code and automated tests written by business analysts and QA testers to cover every desired behavior.

Remember our discussion of tests in Chapter 12, "Harm." Those tests are the scientific *proof* that everything works as it is supposed to. And if everything works as it is supposed to, the next step is to deploy to production.

And by the way, that's how you know if your tests are good enough. Your tests are good enough if, when they pass, you feel comfortable deploying. If passing tests don't allow you to deploy, your tests are deficient.

Perhaps you think that deploying every day or even several times per day would lead to chaos. However, the fact that *you* are ready to deploy does not mean that the *business* is ready to deploy. As part of a development team, your standard is to always be ready.

What's more, we want to help the business remove all the impediments to deployment so that the business deployment cycle can be shortened as much as possible. After all, the more ceremony and ritual the business uses for deployment, the more expensive deployment becomes. Any business would like to shed that expense.

The ultimate goal for any business is continuous, safe, and ceremony-free deployment. Deployment should be as close as possible to a nonevent.

Because deployment is often a lot of work, with servers to configure and databases to load, you need to *automate* the deployment procedure. And because deployment scripts are part of the system, you write tests for them.

For many of you, the idea of continuous deployment may be so far from your current process that it's inconceivable. But that doesn't mean there aren't things you can do to shorten the cycle.

And who knows? If you keep shortening the cycle, month after month, year after year, perhaps one day you'll find that you are deploying continuously.

Continuous Build（持续构建）

Clearly, if you are going to deploy in short cycles, you have to be able to build in short cycles. If you are going to deploy continuously, you have to be able to build continuously.

Perhaps some of you have slow build times. If you do, speed them up. Seriously, with the memory and speed of modern systems, there is no excuse for slow builds. None. Speed them up. Consider it a design challenge.

And then get yourself a continuous build tool, such as Jenkins, Buildbot, or Travis, and use it. Make sure that you kick off a build at every push and do what it takes to ensure that the build never fails.

A build failure is a red alert. It's an emergency. If the build fails, I want emails and text messages sent to every team member. I want sirens going off. I want a flashing red light on the CEO's desk. I want everybody to stop whatever they are doing and deal with the emergency.

Keeping the build from failing is not rocket science. You simply run the build, along with all the tests, in your local environment before you push. You push the code only when all the tests pass.

If the build fails after that, you've uncovered some environmental issue that needs to be resolved posthaste.

You never allow the build to go on failing, because if you allow the build to fail, you'll get used to it failing. And if you get used to the failures, you'll start ignoring them. The more you ignore those failures, the more annoying the failure alerts become. And the more tempted you are to turn the failing tests off until you can fix them—later. You know. Later?

And that's when the tests become lies.

With the failing tests removed, the build passes again. And that makes everyone feel good again. But it's a lie.

So, build continuously. And never let the build fail.

Relentless Improvement（持续改进）

> *Promise 5. I will fearlessly and relentlessly improve my creations at every opportunity. I will never degrade them.*

Robert Baden Powell, the father of the Boy Scouts, left a posthumous message exhorting the scouts to leave the world a better place than they found it. It was from this statement that I derived my Boy Scout rule: Always check the code in cleaner than you checked it out.

How? By performing random acts of kindness upon the code every time you check it in.

One of those random acts of kindness is to increase test coverage.

TEST COVERAGE（测试覆盖率）

Do you measure how much of your code is covered by your tests? Do you know what percentage of lines are covered? Do you know what percentage of branches are covered?

There are plenty of tools that can measure coverage for you. For most of us, those tools come as part of our IDE and are trivial to run, so there's really no excuse for not knowing what your coverage numbers are.

What should you do with those numbers? First, let me tell you what *not* to do. Don't turn them into management metrics. Don't fail the build if your test coverage is too low. Test coverage is a complicated concept that should not be used so naively.

Such naive usage sets up perverse incentives to cheat. And it is very easy to cheat test coverage. Remember that coverage tools only measure the amount of code that was *executed*; not the code that was actually tested. This means that you can drive the coverage number very high by pulling the assertions out of your failing tests. And, of course, that makes the metric useless.

The best policy is to use the coverage numbers as a developer tool to help you improve the code. You should work to *meaningfully* drive the coverage toward 100 percent by writing actual tests.

One hundred percent test coverage is always the goal, but it is also an asymptotic goal. Most systems never reach 100 percent, but that should not deter you from constantly increasing the coverage.

That's what you use the coverage numbers for. You use them as a measurement to help you improve, not as a bludgeon with which to punish the team and fail the build.

MUTATION TESTING（突变测试）

One hundred percent test coverage implies that any semantic change to the code should cause a test to fail. TDD is a good discipline to approximate that goal because, if you follow the discipline ruthlessly, every line of code is written to make a failing test pass.

Such ruthlessness, however, is often impractical. Programmers are human, and disciplines are always subject to pragmatics. So, the reality is that even the most assiduous test-driven developer will leave gaps in the test coverage.

Mutations testing is a way to find those gaps, and there are mutation testing tools that can help. A mutation tester runs your test suite and measures the coverage. Then it goes into a loop, mutating your code in some semantic way and then running the test suite with coverage again. The semantic changes are things like changing > to < or == to `!=` or `x=`<something> to `x=null`. Each such semantic change is called a *mutation*.

The tool expects each mutation to fail the tests. Mutations that do not fail the tests are called *surviving mutations*. Clearly, the goal is to ensure that there are no surviving mutations.

Running a mutation test can be a big investment of time. Even relatively small systems can require hours of runtime, so these kinds of tests are best run over weekends or at month's end. However, I have not failed to be impressed by what mutation testing tools can find. It is definitely worth the occasional effort to run them.

SEMANTIC STABILITY（语义稳定性）

The goal of test coverage and mutation testing is to create a test suite that ensures *semantic stability*. The semantics of a system are the required behaviors of that system. A test suite that ensures semantic stability is one that fails whenever a required behavior is broken. We use such test suites to eliminate the fear of refactoring and cleaning the code. Without a semantically stable test suite, the fear of change is often too great.

TDD gives us a good start on a semantically stable test suite, but it is not sufficient for full semantic stability. Coverage, mutation testing, and acceptance testing should be used to improve the semantic stability toward completeness.

Cleaning（清理）

Perhaps the most effective of the random acts of kindness that will improve the code is simple cleaning—refactoring with the goal of improvement.

What kinds of improvements can be made? There is, of course, the obvious elimination of code smells. But I often clean code even when it isn't smelly.

I make tiny little improvements in the names, in the structure, in the organization. These changes might not be noticed by anyone else. Some folks might even think they make the code less clean. But my goal is not simply the state of the code. By doing little minor cleanings, I *learn* about the code. I become more familiar and more comfortable with it. Perhaps my cleaning did not actually improve the code in any objective sense, but it improved my understanding of and my facility with that code. The cleaning improved *me* as a developer of that code.

The cleaning provides another benefit that should not be understated. By cleaning the code, even in minor ways, I am *flexing* that code. And one of the best ways to ensure that code stays flexible is to regularly flex it. Every little bit of cleaning I do is actually a test of the code's flexibility. If I find a small cleanup to be a bit difficult, I have detected an area of inflexibility that I can now correct.

Remember, software is supposed to be soft. How do you know that it is soft? By testing that softness on a regular basis. By doing little cleanups and little improvements and *feeling* how easy or difficult those changes are to make.

CREATIONS（创造）

The word used in Promise 5 is *creations*. In this chapter, I have focused mostly on code, but code is not the only thing that programmers create. We create designs and documents and schedules and plans. All of these artifacts are creations that should be continuously improved.

We are human beings. Human beings make things better with time. We constantly improve everything we work on.

MAINTAIN HIGH PRODUCTIVITY（保持高生产力）

> *Promise 6. I will do all that I can to keep the productivity of myself and others as high as possible. I will do nothing that decreases that productivity.*

Productivity. That's quite a topic, isn't it? How often do you feel that it's the only thing that matters at your job? If you think about it, productivity is what this book and all my books on software are about.

They're about how to go fast.

And what we've learned over the last seven decades of software is that the way you go fast is to go well.

The *only* way to go fast is to go well.

So, you keep your code clean. You keep your designs clean. You write semantically stable tests and keep your coverage high. You know and use appropriate design patterns. You keep your methods small and your names precise.

But those are all *indirect* methods of achieving productivity. Here, we're going to talk about much more direct ways to keep productivity high.

1. Viscosity—keeping your development environment efficient
2. Distractions—dealing with every day business and personal life

3. Time management—effectively separating productive time from all the other junk you have to do

Viscosity（拖慢速度的因素）

Programmers are often very myopic when it comes to productivity. They view the primary component to productivity as their ability to write code quickly.

But the writing of code is a very small part of the overall process. If you made the writing of code *infinitely* fast, it would increase overall productivity only by a small amount.

That's because there's a lot more to the software process than just writing code. There's at least

- Building
- Testing
- Debugging
- Deploying

And that doesn't count the requirements, the analysis, the design, the meetings, the research, the infrastructure, the tooling, and all the other stuff that goes into a software project.

So, although it is important to be able to write code efficiently, it's not even close to the biggest part of the problem.

Let's tackle some of the other issues one at a time.

Building

If it takes you 30 minutes to build after a 5-minute edit, then you can't be very productive, can you?

There is no reason, in the second and subsequent decades of the twenty-first century, that builds should take more than a minute or two.

Before you object to this, think about it. How could you speed the build? Are you utterly certain, in this age of cloud computing, that there is no way to dramatically speed up your build? Find whatever is causing the build to be slow, and fix it. Consider it a design challenge.

Testing

Is it your tests that are slowing the build? Same answer. Speed up your tests.

Here, look at it this way. My poor little laptop has four cores running at a clock rate of 2.8GHz. That means it can execute around *10 billion instructions per second.*

Do you even have 10 billion instructions in your whole system? If not, then you should be able to test your whole system in less than a second.

Unless, of course, you are executing some of those instructions more than once. For example, how many times do you have to test login to know that it works? Generally speaking, once should be sufficient. How many of your tests go through the login process? Any more than one would be a waste!

If login is required before each test, then during tests, you should short-circuit the login process. Use one of the mocking patterns. Or, if you must, remove the login process from systems built for testing.

The point is, don't tolerate repetition like that in your tests. It can make them horrifically slow.

As another example, how many times do your tests walk through the navigation and menu structure of the user interface? How many tests start at the top and then walk through a long chain of links to finally get the system into the state where the test can be run?

Any more than once per navigation pathway would be a waste! So, build a special testing API that allows the tests to quickly force the system into the state you need, without logging in and without navigating.

How many times do you have to execute a query to know that it works? Once! So, mock out your databases for the majority of your tests. Don't allow the same queries to be executed over and over and over again.

Peripheral devices are slow. Disks are slow. Web sockets are slow. UI screens are slow. Don't let slow things slow down your tests. Mock them out. Bypass them. Get them out of the critical path of your tests.

Don't tolerate slow tests. Keep your tests running fast!

Debugging

Does it take a long time to debug things? Why? Why is debugging slow?

You are using TDD to write unit tests, aren't you? You are writing acceptance tests too, right? And you are measuring test coverage with a good coverage analysis tool, right? And you are periodically proving that your tests are semantically stable by using a mutation tester, right?

If you are doing all those things or even just *some* of those things, your debug time can be reduced to insignificance.

Deployment

Does deployment take forever? Why? I mean, you *are* using deployment scripts, right? You aren't deploying manually, are you?

Remember, you are a programmer. Deployment is a procedure—automate it! And write tests for that procedure too!

You should be able to deploy your system, every time, with a single click.

Managing Distractions（解决注意力分散问题）

One of the most pernicious destroyers of productivity is distraction from the job. There are many different kinds of distractions. It is important for you to know how to recognize them and defend against them.

Meetings

Are you slowed down by meetings?

I have a very simple rule for dealing with meetings. It goes like this:

> *When the meeting gets boring, leave.*

You should be polite about it. Wait a few minutes for a lull in the conversation, and then tell the participants that you believe your input is no longer required and ask them if they would mind if you returned to the rather large amount of work you have to do.

Never be afraid of leaving a meeting. If you don't figure out how to leave, then some meetings will keep you forever.

You would also be wise to decline most meeting invitations. The best way to avoid getting caught in long, boring meetings is to politely refuse the invitation in the first place. Don't be seduced by the fear of missing out. If you are truly needed, they'll come get you.

When someone invites you to a meeting, make sure they've convinced you that you really need to go. Make sure they understand that you can only afford a few minutes and that you are likely to leave before the meeting is over.

And make sure you sit close to the door.

If you are a group leader or a manager, remember that one of your primary duties is to defend your team's productivity by keeping them out of meetings.

Music

I used to code to music long, long ago. But I found that listening to music impedes my concentration. Over time, I realized that listening to music only *feels* like it helps me concentrate, but actually, it divides my attention.

One day, while looking over some year-old code, I realized that my code was suffering under the lash of the music. There, scattered through the code in a series of comments, were the lyrics to the song I had been listening to.

Since then, I've stopped listening to music while I code, and I've found I am much happier with the code I write and with the attention to detail that I can give it.

Programming is the act of arranging elements of procedure through sequence, selection, and iteration. Music is composed of tonal and rhythmic elements arranged through sequence, selection, and iteration. Could it be that listening to music uses the same parts of your brain that programming uses, thereby consuming part of your programming ability? That's my theory, and I'm sticking to it.

You will have to work this out for yourself. Maybe the music really does help you. But maybe it doesn't. I'd advise you to try coding without music for a week, and see if you don't end up producing more and better code.

Mood

It's important to realize that being productive requires that you become skilled at managing your emotional state. Emotional stress can kill your ability to code. It can break your concentration and keep you in a perpetually distracted state of mind.

For example, have you ever noticed that you can't code after a huge fight with your significant other? Oh, maybe you type a few random characters in your IDE, but they don't amount to much. Perhaps you pretend to be productive by hanging out in some boring meeting that you don't have to pay much attention to.

Here's what I've found works best to restore your productivity.

Act. Act on the root of the emotion. Don't try to code. Don't try to cover the feelings with music or meetings. It won't work. Act to resolve the emotion.

If you find yourself at work, too sad or depressed to code because of a fight with your significant other, then call them to try to resolve the issue. Even if you don't actually get the issue resolved, you'll find that the action to attempt a resolution will sometimes clear your mind well enough to code.

You don't actually have to solve the problem. All you have to do is convince yourself that you've taken enough appropriate action. I usually find that's enough to let me redirect my thoughts to the code I have to write.

The Flow

There's an altered state of mind that many programmers enjoy. It's that hyperfocused, tunnel-vision state in which the code seems to pour out of every orifice of your body. It can make you feel superhuman.

Despite the euphoric sensation, I've found, over the years, that the code I produce in that altered state tends to be pretty bad. The code is not nearly as well considered as code I write in a normal state of attention and focus. So, nowadays, I resist getting into the flow. Pairing is a very good way to stay out of the flow. The very fact that you must communicate and collaborate with someone else seems to interfere with the flow.

Avoiding music also helps me stay out of the flow because it allows the actual environment to keep me grounded in the real world.

If I find that I'm starting to hyperfocus, I break away and do something else for a while.

Time Management（时间管理）

One of the most important ways to manage distraction is to employ a time management discipline. The one I like best is *the Pomodoro Technique*.[2]

2. Francesco Cirillo, *The Pomodoro Technique: The Life-Changing Time-Management System* (Virgin Books, 2018).

Pomodoro is Italian for “tomato.” English teams tend to use the word *tomato* instead. But you’ll have better luck with Google if you search for the Pomodoro Technique.

The aim of the technique is to help you manage your time and focus during a regular workday. It doesn’t concern itself with anything beyond that.

At its core, the idea is quite simple. Before you begin to work, you set a timer (traditionally a kitchen timer in the shape of a tomato) for 25 minutes.

Next, you work. And you work until the timer rings.

Then you break for 5 minutes, clearing your mind and body.

Then you start again. Set the timer for 25 minutes, work until the timer rings, and then break for 5 minutes. You do this over and over.

There’s nothing magical about 25 minutes. I’d say anything between 15 and 45 minutes is reasonable. But once you choose a time, stick with that time. Don’t change the size of your tomatoes!

Of course, if I were 30 seconds away from getting a test to pass when the timer rang, I’d finish the test. On the other hand, maintaining the discipline is important. I wouldn’t go more than a minute beyond.

So far, this sounds mundane, but handling interruptions, such as phone calls, is where this technique shines. The rule is to *defend the tomato!*

Tell whoever is trying to interrupt you that you’ll get back to them within 25 minutes—or whatever the length of your tomato is. Dispatch the interruption as quickly as possible, and then return to work.

Then, after your break, handle the interruption.

This means that the time between tomatoes will sometimes get pretty long, because people who interrupt you often require a lot of time.

Again, that's the beauty of this technique. At the end of the day, you count the number of tomatoes you completed, and that gives you a measure of your productivity.

Once you've gotten good at breaking your day up into tomatoes like this and defending the tomatoes from interruptions, then you can start planning your day by allocating tomatoes to it. You may even begin to estimate your tasks in terms of tomatoes and plan your meetings and lunches around them.

14 TEAMWORK

（团队合作）

The remaining promises of the oath reflect a commitment to the team.

Work as a Team（组团工作）

Promise 7. I will continuously ensure that others can cover for me and that I can cover for them.

Segregation of knowledge into silos is extremely detrimental to a team and to an organization. The loss of an individual can mean the loss of an entire segment of knowledge. It can paralyze the team and the organization. It also means that the individuals on the team don't have sufficient context to understand each other. Often, they wind up talking past each other.

The cure for this problem is to spread knowledge through the team. Make sure each team member knows a lot about the work that other team members are performing.

The best way to spread that knowledge is to work together—to pair, or mob.

The truth is that there's hardly any better way to improve the productivity of a team than to practice collaborative programming. A team that knows the deep connections among the work that's being done can't help but be much more productive than a group of silos.

Open/Virtual Office（开放式 / 虚拟办公室）

It is also important that the members of the team see and interact with each other very frequently. The best way to achieve this is to put them into a room together.

In the early 2,000s, I owned a company that helped organizations adopt Agile development. We would send a group of instructors and coaches to those companies and guide them through the change. Before each engagement began, we told the managers to rearrange the office space so that the teams we would be coaching worked in their own team rooms. It happened more than once that, before we arrived to begin coaching, the managers told us that the teams were already much more productive just because they were working together in the same room.

I'm writing this paragraph in the first quarter of 2021. The COVID-19 pandemic is beginning to wane, vaccines are being rapidly distributed (I will get my second shot today), and we are all looking forward to a return to normal life. But the pandemic will leave, in its wake, a large number of software teams who work remotely.

Working remotely can never be quite as productive as working together in the same room. Even with the best electronic help, seeing each other on screens is just not as good as seeing each other in person. Still, the electronic systems for collaboration are very good nowadays. So, if you are working remotely, *use them.*

Create a virtual team room. Keep everyone's face in view. Keep everyone's audio channel as open as feasible. Your goal is to create the illusion of a team room, with everyone in it working together.

Pairing and mobbing enjoy a lot of electronic support nowadays. It is relatively easy to share screens and program together across distances. While you do that, keep the faces and the audio up and running. You want to be able to *see* each other while you are collaborating on code.

Remote teams should try, as hard as they can, to maintain the same working hours. This is very difficult when there is a huge East–West distribution of programmers. Try to keep the number of time zones in each team as small as possible, and try very hard to have at least six contiguous hours per day when everyone can be together in the virtual team room.

Have you ever noticed how easy it is to yell at another driver while driving your car? This is the windshield effect. When you are sitting behind a windshield, it's easy to see other people as fools, imbeciles, and even enemies. It is easy to dehumanize them. This effect happens, to a lesser degree, behind computer screens.

To avoid this effect, teams should get together in the same physical room several times per year. I recommend one week each quarter. This will help the team congeal and maintain itself as a team. It is very hard to fall into the windshield trap with someone you ate lunch with and physically collaborated with two weeks ago.

Estimate Honestly and Fairly（诚实和合理地预估）

Promise 8. I will produce estimates that are honest in both magnitude and precision. I will not make promises without reasonable certainty.

In this section, we're going to talk about estimating projects and large tasks—things that take many days or weeks to accomplish. The estimation of small tasks and stories is described in *Clean Agile*.[1]

Knowing how to estimate is an essential skill for every software developer and one that most of us are very, very bad at. The skill is essential because every business needs to know, roughly, how much something is going to cost before they commit resources to it.

Unfortunately, our failure to understand what estimates actually are and how to create them has led to an almost catastrophic loss of trust between programmers and business.

The landscape is littered with billions of dollars in software failures. Often, those failures are due to poor estimation. It is not uncommon for estimates to be off by a factor of 2, 3, even 4 and 5. But why? Why are estimates so hard to get right?

Mostly it's because we don't understand what estimates actually are and how to create them. You see, in order for estimates to be useful, they must be honest: They must be honestly accurate, and they must be honestly precise. But most estimates are neither. Indeed, most estimates are lies.

1. Robert C Martin, *Clean Agile: Back to Basics* (Pearson, 2020).

Lies（谎言）

Most estimates are lies because most estimates are constructed backward from a known end date.

Consider HealthCare.gov, for example. The president of the United States signed a bill into law that mandated a specific date when that software system was to be turned on.

The depth of that illogic is nausea inducing. I mean, how absurd. Nobody was asked to estimate the end date; they were just told what the end date had to be—by law!

So, of course, all estimates associated with that mandated date were lies. How could they be anything else?

It reminds me of a team I was consulting for about twenty years ago. I remember being in the project room with them when the project manager walked in. He was a young fellow, perhaps twenty-five. He'd just returned from a meeting with his boss. He was visibly agitated. He told the team how important the end date was. He said, "We really have to make that date. I mean, we *really* have to make that date."

Of course, the rest of the team just rolled their eyes and shook their heads. The need to make the date was not a solution for making the date. The young manager offered no solution.

Estimates, in an environment like that, are just lies that support the plan.

And that reminds me of another client of mine who had a huge software production plan on the wall—full of circles and arrows and labels and tasks. The programmers referred to it as *the laugh track*.

In this section, we're going to talk about real, worthwhile, honest, accurate, and precise estimates. The kind of estimates that professionals create.

HONESTY, ACCURACY, PRECISION（诚实、准确、精确）

The most important aspect of an estimate is honesty. Estimates don't do anybody any good unless they are honest.

Me: So, let me ask you. What is the most honest estimate you can give?

Programmer: *Um, I don't know.*

Me: *Right.*

Programmer: *Right what?*

Me: *I don't know.*

Programmer: *Wait. You asked me for the most honest estimate.*

Me: *Right.*

Programmer: *And I said I don't know.*

Me: *Right.*

Programmer: *So, what it is?*

Me: *I don't know.*

Programmer: *Well, then, how do you expect me to know?*

Me: *You already said it?*

Programmer: *Said what?*

Me: I don't know.

The most honest estimate you can give is "I don't know." But that estimate is not particularly accurate or precise. After all, you do know *something* about the estimate. The challenge is to quantify what you do and don't know.

First, your estimate must be accurate. That doesn't mean you give a firm date—you don't dare be that precise. It just means that you name a range of dates that you feel confident in.

So, for example, sometime between now and ten years from now is a pretty accurate estimate for how long it would take you to write a hello world program. But it lacks precision.

On the other hand, yesterday at 2:15 a.m. is a very precise estimate, but it's probably not very accurate if you haven't started yet.

Do you see the difference? When you give an estimate, you want it to be honest both in accuracy and in precision. To be accurate, you name a range of dates within which you are confident. To be precise, you narrow that range up to the level of your confidence.

And for both of these operations, brutal honesty is the only option.

To be honest about these things, you have to have some idea of how wrong you can be. So let me tell you two stories about how wrong I once was.

Story 1: Vectors（故事1：载体）

The year was 1978. I was working at a company named Teradyne, in Deerfield, Illinois. We built automated test equipment for the telephone company.

I was a young programmer, 26 years old. And I was working on the firmware for an embedded measurement device that bolted into racks in telephone central offices. This device was called a COLT—central office line tester.

The processor in the COLT was an Intel 8085—an early 8-bit microprocessor. We had 32K of solid-state RAM and another 32K of ROM. The ROM was based on the Intel 2708 chip, which stored 1K × 8, so we used 32 of those chips.

Those chips were plugged into sockets on our memory boards. Each board could hold 12 chips, so we used three boards.

The software was written in 8085 assembler. The source code was held in a set of source files that were compiled as a single unit. The output of the compiler was a single binary file somewhat less than 32K in length.

We took that file and cut it up into 32 chunks of 1K each. Each 1K chunk was then burned onto one of the ROM chips, which were then inserted into the sockets on the ROM boards.

As you can imagine, you had to get the right chip into the right socket on the right board. We were very careful to label them.

We sold hundreds of these devices. They were installed in telephone central offices all over the country and—indeed—the world.

What do you think happened when we changed that program? Just a one-line change?

If we added or removed a line, then all the addresses of all the subroutines after that point changed. And because those subroutines were called by other routines earlier in the code, every chip was affected. We had to reburn all 32 chips even for a one-line change!

It was a nightmare. We had to burn hundreds of sets of chips and ship them to all the field service reps all around the world. Then those reps would have to drive hundreds of miles to get to all the central offices in their district. They'd have to open our units, pull out all the memory boards, remove all 32 old chips, insert the 32 new chips, and reinsert the boards.

Now, I don't know if you know this, but the act of removing and inserting a chip into a socket is not entirely reliable. The little pins on the chips tend to bend and break in frustratingly silent ways. Therefore, the poor field service folks had to have lots of spares of each of the 32 chips and suffer through the inevitable debugging by removing and reinserting chips until they could get a unit to work.

My boss came to me one day and told me that we had to solve this problem by making each chip independently deployable. He didn't use those words, of course, but that was the intent. Each chip needed to be turned into an independently compilable and deployable unit. This would allow us to make changes to the program without forcing all 32 chips to be reburned. Indeed, in most cases, we could simply redeploy a single chip—the chip that was changed.

I won't bore you with the details of the implementation. Suffice it to say that it involved vector tables, indirect calls, and the partitioning of the program into independent chunks of less than 1K each.[2]

My boss and I talked through the strategy, and then he asked me how long it would take me to get this done.

I told him two weeks.

But I wasn't done in two weeks. I wasn't done in four weeks. Nor was I done in six, eight, or ten weeks. The job took me twelve weeks to complete—it was a lot more complicated than I had anticipated.

Consequently, I was off by a factor of 6. Six!

Fortunately, my boss didn't get mad. He saw me working on it every day. He got regular status updates from me. He understood the complexities I was dealing with.

But still. Six? How could I have been so wrong?

Story 2: pCCU（故事2：pCCU）

Then there was that time, in the early 1980s, when I had to work a miracle.

2. That is, each chip was turned into a polymorphic object.

You see, we had promised a new product to our customer. It was called CCU-CMU.

Copper is a precious metal. It's rare and expensive. The phone company decided to harvest the huge network of copper wires that it had installed all over the country during the last century. The strategy was to replace those wires with a much cheaper high-bandwidth network of coaxial cable and fiber carrying digital signals. This was known as *digital switching*.

The CCU-CMU was a complete re-architecture of our measurement technology that fit within the new digital switching architecture of the phone company.

Now, we had promised the CCU-CMU to the phone company a couple of years before. We knew it was going to take us a man-year or so to build the software. But then we just never quite got around to building it.

You know how it goes. The phone company delayed their deployment, so we delayed our development. There were always lots of other, more urgent issues to deal with.

So, one day, my boss calls me into his office and says that they had forgotten about one small customer who had already installed an early digital switch. That customer was now expecting a CCU/CMU within the next month—as promised.

Now I had to create a man-year of software in less than a month.

I told my boss that this was impossible. There was no way that I could get a fully functioning CCU/CMU built in one month.

He looked at me with a sneaky grin and said that there was a way to cheat.

You see, this was a very small customer. Their installation was literally the smallest possible configuration for a digital switch. What's more, the configuration of their equipment just happened—just happened—to eliminate virtually all of the complexity that the CCU/CMU solved.

Long story short—I got a special-purpose, one-of-a-kind unit up and running for the customer in two weeks. We called it the pCCU.

The Lesson（教训）

Those two stories are examples of the huge range that estimates can have. On the one hand, I underestimated the vectoring of the chips by a factor of six. On the other, we found a solution to the CCU/CMU in one-twentieth the expected time.

This is where honesty comes in. Because, honestly, when things go wrong, they can go very, very wrong. And when things go right, they can sometimes go very, very right.

This makes estimating, one hell of a challenge.

Accuracy（准确度）

It should be clear by now that an estimate for a project *cannot* be a date. A single date is far too precise for a process that can be off by as much as a factor of 6, or even 20.

Estimates are not dates. Estimates are ranges. Estimates are *probability distributions*.

Probability distributions have a mean and a width—sometimes called the *standard deviation* or the *sigma*. We need to be able to express our estimates as both the mean and the sigma.

Let's first look at the mean.

Finding the expected mean completion time for a complex task is a matter of adding up all the mean completion times for all of the subtasks. And, of course, this is recursive. The subtasks can be estimated by adding up all the times for the sub-subtasks. This creates a tree of tasks that is often called a work breakdown structure (WBS).

Now, this is all well and good. The problem, however, is that we are not very good at identifying all the subtasks and sub-subtasks, and sub-sub-subtasks. Generally, we miss a few. Like, maybe, half.

We compensate for this by multiplying the sum by two. Or sometimes three. Or maybe even more.

Kirk: *How much refit time until we can take her out again?*

Scotty: *Eight weeks, sir. But you don't have eight weeks, so I'll do it for you in two.*

Kirk: *Mr. Scott, have you always multiplied your repair estimates by a factor of four?*

Scotty: Certainly, sir! How else can I keep my reputation as a miracle worker?[3]

Now this fudge factor of 2, or 3, or even 4, sounds like cheating. And, of course, it is. But so is the very act of estimating.

There is only one real way to determine how long something is going to take, and that's by doing it. Any other mechanism is cheating.

So, face it, we're going to cheat. We're going to do the WBS and then multiply by some *F*, where *F* is between 2 and 4, depending on your confidence and productivity. That will give us our mean time to completion.

Managers are going to ask you how you came up with your estimates, and you're going to have to tell them. And when you tell them about that fudge factor, they're going to ask you to reduce it by spending more time on the WBS.

3. *Star Trek II: The Wrath of Khan,* directed by Nicholas Meyer (Paramount Pictures, 1982).

This is perfectly fair, and you should be willing to comply. However, you should also warn them that the cost of developing a complete WBS is equivalent to the cost of the task itself. Indeed, by the time you *have* developed the complete WBS, you will also have completed the project, because the only way to truly enumerate all the tasks is by executing the tasks you know about in order to discover the rest—recursively.

So, make sure you put your estimation effort into a timebox, and let your managers know that getting better refinement on the fudge factor is going to be very expensive.

There are many techniques for estimating the subtasks at the leaves of the WBS tree. You could use function points or a similar complexity measure. But I've always found that these tasks are best estimated by raw gut feel.

Typically, I do this by comparing the tasks to some other tasks that I've already completed. If I think it's twice as hard, I multiply the time by two.

Once you've estimated all the leaves of the tree, you just sum the whole tree up to get the mean for the project.

And don't worry overmuch about dependencies. Software is a funny material. Even though A depends on B, B often does not have to be done before A. You *can*, in fact, implement logout before you implement login.

Precision（精确度）

Every estimate is wrong. That's why we call it an estimate. An estimate that is correct is not an estimate at all—it's a fact.

But even though the estimate is wrong, it may not be all that wrong. Part of the job of estimation, then, is to estimate how wrong the estimate is.

My favorite technique for estimating how wrong an estimate is, is to estimate three numbers: the best case, the worst case, and the normal case.

The normal case is how long you think the task would take you if the average number of things go wrong—if things go the way they *usually* do. Think of it like a gut-level call. The normal case is the estimate you would give if you were being realistic.

The strict definition of a normal estimate is one that has a 50 percent chance of being too short or too long. In other words, you should miss half of your normal estimates.

The worst-case estimate is the Murphy's law estimate. It assumes that anything that can go wrong will go wrong. It is deeply pessimistic. The worst-case estimate has a 95 percent chance of being too long. In other words, you would only miss this estimate 1 in 20 times.

The best-case estimate is when everything possible goes right. You eat the right breakfast cereal every morning. When you get into work, your coworkers are all polite and friendly. There are no disasters in the field, no meetings, no telephone calls, no distractions.

Your chances of hitting the best-case estimate are 5 percent: 1 in 20.

Okay, so now we have three numbers: the best case, which has a 5 percent chance of success; the normal case, which has a 50 percent chance of success; and the worst case, which has a 95 percent chance of success. This represents a normal curve—a probability distribution. It is this probability distribution that is your actual estimate.

Notice that this is not a date. We don't know the date. We don't know when we are really going to be done. All we really have is a crude idea of the probabilities.

Without certain knowledge, probabilities are the only logical way to estimate.

If your estimate is a date, you are really making a commitment, not an estimate. And if you make a commitment, you *must* succeed.

Sometimes you have to make commitments. But the thing about commitments is that you absolutely must succeed. You must never promise to make a date that you aren't sure you can make. To do so would be deeply dishonest.

So, if you don't know—and I mean *know*—that you can make a certain date, then you don't offer that date as an estimate. You offer a range of dates instead. Offering a range of dates with probabilities is much more honest.

AGGREGATION（汇总）

Okay, let's say that we've got a whole project full of tasks that have been described in terms of Best (B), Normal (N), and Worst (W) case estimates. How do we aggregate them all into a single estimate for the whole project?

We simply represent the probability of each task and then accumulate those probabilities using standard statistical methods.

The first thing we want to do is represent each task in terms of its expected completion time and the standard deviation.

Now remember, 6 standard deviations (3 on each side of the mean) corresponds to a probability of better than 99 percent. So, we're going to set our standard deviation, our sigma, to Worst minus Best over 6.

The expected completion time (mu) is a bit trickier. Notice that N is probably not equal to (W-B), the midpoint. Indeed, the midpoint is probably well past N because it is much more likely for a project to take more time than we think than less time. So, on average, when will this task be done? What is the *expected* completion time?

It's probably best to use a weighted average like this: mu = (2N + (B + W)/2)/3.

Now we have calculated the mu and sigma for a set of tasks. The expected completion time for the whole project is just the sum of all the mus.
The sigma for the project is the square root of the sum of the squares of all the sigmas.

This is just basic statistical mathematics.

What I've just described is the estimation procedure invented back in the late 1950s to manage the Polaris Fleet Ballistic Missile program. It has been used successfully on many thousands of projects since.

It is called PERT—the program evaluation and review technique.

HONESTY（诚实）

We started with honesty. Then we talked about accuracy, and we talked about precision. Now it's time to come back to honesty.

The kind of estimating we are talking about here is intrinsically honest. It is a way of communicating, to those who need to know, the level of your uncertainty.

This is honest because you truly are uncertain. And those with the responsibility to manage the project must be aware of the risks that they are taking so that they can manage those risks.

But uncertainty is something that people don't like. Your customers and managers will almost certainly press you to be more certain.

We've already talked about the cost of increasing certainty. The only way to truly increase certainty is to do parts of the project. You can only get perfect certainty if you do the entire project. Therefore, part of what you have to tell your customers and managers is the cost of increasing certainty.

Sometimes, however, your superiors may ask you to increase certainty using a different tactic. They may ask you to commit.

You need to recognize this for what it is. They are trying to manage their risk by putting it onto you. By asking you to commit, they are asking you to take on the risk that it is their job to manage.

Now, there's nothing wrong with this. Managers have a perfect right to do it. And there are many situations in which you should comply. But—and I stress this—only if you are reasonably certain you *can* comply.

If your boss comes to you and asks if you can get something done by Friday, you should think very hard about whether that is reasonable. And if it is reasonable and probable, then, by all means, say yes!

But under no circumstances should you say yes if you are not sure.

If you are not sure, then you *must* say NO and then describe your uncertainty as we've described. It is perfectly okay to say, "I can't promise Friday. It might take as long as the following Wednesday."

In fact, it's absolutely critical that you say no to commitments that you are not sure of, because if you say yes, you set up a long domino chain of failures for you, your boss, and many others. They'll be counting on you, and you'll let them down.

So, when you are asked to commit and you can, then say yes. But if you can't, then say no and describe your uncertainty.

Be willing to discuss options and workarounds. Be willing to hunt for ways to say yes. Never be eager to say no. But also, never be afraid to say no.

You see, you were hired for your ability to say no. Anybody can say yes. But only people with skill and knowledge know when and how to say no.

One of the prime values you bring to the organization is your ability to know when the answer must be no. By saying no at those times, you will save your company untold grief and money.

One last thing. Often, managers will try to cajole you into committing—into saying yes. Watch out for this.

They might tell you that you aren't being a team player or that other people have more commitment than you do. Don't be fooled by those games.

Be willing to work with them to find solutions, but don't let them bully you into saying yes when you know you shouldn't.

And be very careful with the word *try*. Your boss might say something reasonable, like "Well, will you at least try?"

The answer to this question is

> *NO! I am already trying. How dare you suggest that I am not? I am trying as hard as I can, and there is no way I can try harder. There are no magic beans in my pocket with which I can work miracles.*

You might not want to use those exact words, but that's exactly what you should be thinking.

And remember this. If you say, "Yes, I'll try," then you are lying. Because you have no idea how you are going to succeed. You don't have any plan to change your behavior. You said yes just to get rid of the manager. And that is fundamentally dishonest.

Respect（尊重）

> *Promise 9. I will respect my fellow programmers for their ethics, standards, disciplines, and skill. No other attribute or characteristic will be a factor in my regard for my fellow programmers.*

We, software professionals, accept the weighty burden of our craft. We brave folks are men and women, straight and gay, black, brown, yellow, and white, republicans, democrats, religious believers, and atheists. We are humans, in all the many forms and varieties that humans come in. We are a community of mutual respect.

The only qualifications for entry into our community and for receiving the acceptance and the respect of each and every member of that community are the skills, disciplines, standards, and ethics of our profession. No other human attribute is worthy of consideration. No discrimination on any other basis can be tolerated.

'Nuff said.

Never Stop Learning（永不停止学习）

Promise 10. I will never stop learning and improving my craft.

A programmer never stops learning.

I'm sure you've heard it said that you should learn a new language every year. Well, you should. A good programmer should know a dozen or so languages.

And not just a dozen varieties of the same language. Not just C, C++, Java, and C#. Rather, you should know languages from many different families.

You should know a statically typed language like Java or C#. You should know a procedural language like C or Pascal. You should know a logic language like Prolog. You should know a stack language like Forth. You should know a dynamically typed language like Ruby. You should know a functional language like Clojure or Haskell.

You should also know several different frameworks, several different design methodologies, and several different development processes. I don't mean to say you should be an expert in all these things, but you should make a point to expose yourself to them at significantly more than a cursory level.

The list of things you should similarly expose yourself to is virtually endless. Our industry has experienced rapid change over the decades, and that change is likely to continue for some time. You have to keep up with it.

And that means you have to keep on learning. Keep reading books and blogs. Keep watching videos. Keep going to conferences and user groups. Keep going to training courses. Keep learning.

Pay attention to the treasured works of the past. The books written in the 1960s, 1970s, and 1980s are wonderful sources of insight and information. Don't fall into the trap of thinking that all that old stuff is out of date. There is not much, in our industry, that actually goes out of date. Respect the effort and accomplishments of those who came before you, and study their advice and conclusions.

And don't fall into the trap of thinking that it is your employer's job to train you. This is *your* career—you have to take responsibility for it. It is your job to learn. It is your job to figure out what you should be learning.

If you are lucky enough to work for a company that will buy you books and send you to conferences and training classes, then take full advantage of those opportunities. If not, then pay for those books, conferences, and courses yourself.

And plan to spend some time at this. Time every week. You owe your employer 35 to 40 hours per week. You owe your career another 10 to 20.

That's what professionals do. Professionals put in the time to groom and maintain their careers. And that means you should be working 50 to 60 hours per week total. Mostly at work, but a lot at home too.